John Hutton Balfour

**The Plants of the Bible**

John Hutton Balfour

**The Plants of the Bible**

ISBN/EAN: 9783337096502

Printed in Europe, USA, Canada, Australia, Japan

Cover: Foto ©berggeist007 / pixelio.de

More available books at **www.hansebooks.com**

ARBOUR COVERED WITH A GOURD.

Page 215.

# The Plants of the Bible.

THE GRAPE VINE.
Page 103.

**THOMAS NELSON AND SONS,**
LONDON, EDINBURGH, AND NEW YORK.

# THE PLANTS OF THE BIBLE.

*By*

*JOHN HUTTON BALFOUR, M.A., M.D.,*
*F.R.S.S.L. & E., F.L.S., F.R.C.S.E.*

New and Enlarged Edition.

London:
T. NELSON AND SONS, PATERNOSTER ROW.
EDINBURGH; AND NEW YORK.

1885.

# Author's Preface.

EVERYTHING mentioned in the Bible is worthy of our attentive consideration. The very words of the original text, written by the inspiration of the Holy Spirit, call for diligent study; and the more we examine them in dependence on the aid of that Spirit, the more light do we find shining upon them. The student of God's Word cannot search too deeply or too minutely into its hidden treasures. The most learned linguist finds here ample scope for all his lore, and the accomplished naturalist may bring to bear upon this work all the resources of science.

In the Sacred Writings there are frequent allusions made to the Vegetable Kingdom. Our blessed Saviour drew beautiful illustrations from plants, and he calls upon us to consider the lilies of the field. While plants, like the other works of the Almighty Creator, are well worthy of study, they are especially so when we view them in connection with Scripture. In order to see fully the lesson which is to be taught, it is necessary that we should be acquainted with the plant to which reference is made. Want of knowledge in this respect has hid much of the beauty and force of many a parable.

At the time when our excellent English version was made, there was comparatively little known in regard to the plants of Palestine, and hence the meaning of the Hebrew and Greek names was often doubtfully given. As the science of Botany has advanced, and more particularly as the knowledge of the Flora of the East has increased, additional light has been thrown on the plants noticed in the Bible. Celsius, Rosenmüller, Royle, and many others, have done much to elucidate Scripture Botany; and although there are still many difficulties in the way of a complete Bible Flora, still there has been a great advance in this department of Biblical learning. It has been thought that such a work as the present might be useful in calling attention to this important subject, and in inducing those who may visit Palestine to turn their powers of observation to useful account. It is to be regretted that, of the numerous visitors to the Holy Land in recent times, few have turned their thoughts in this direction, and that thus many valuable opportunities for acquiring botanical information have been lost. The Botany of the Bible can be fully worked out only by those who travel in Eastern countries, and who are acquainted with Hebrew, Syriac, Arabic, and other cognate languages. A great deal of valuable information may be gathered on the spot, which cannot be otherwise obtained. Let us hope that, ere long, travellers will have greater facilities for prosecuting with safety their researches in that interesting, although now deserted, land; and that some botanist may soon arise who will be able to write with scientific accuracy on all the Scripture plants, from the Cedar on Lebanon even to the Hyssop that groweth out of the wall.

### *NOTE BY THE PUBLISHERS.*

Dr. Balfour died while this book was passing through the press. The whole of it had passed under his eye in proof sheet before his death. To the last he continued to take great interest in it, as representing a union of the two departments of study which had chiefly occupied his life. The little book will have a special interest for his friends, as the revision of these pages was the last literary work in which their author engaged.

PARKSIDE, EDINBURGH,
*January 1885.*

# Contents.

| | |
|---|---|
| ALMOND-TREE (*Amygdalus communis; Prunus amygdalus*) | 9 |
| BOX-TREE (*Boxus sempervirens*) | 14 |
| BAY-TREE (*Laurus nobilis*) | 18 |
| CEDAR-TREE OF LEBANON (*Cedrus Libani*) | 21 |
| HEATH-TREE—SAVIN (*Juniperus Sabina*) | 28 |
| CINNAMON-TREE AND CASSIA-TREE (*Cinnamomum zeylanicum; and C. Cassia*) | 30 |
| FIR-TREE (*Cupressus sempervirens*) | 34 |
| FIG-TREE (*Ficus Carica*) | 40 |
| HYSSOP (*Capparis spinosa; Caper plant and its variety, Capparis ægyptiaca*) | 44 |
| ASPEN, OR TREMBLING POPLAR (*Populus tremula*) | 48 |
| OAK-TREE (*Quercus Ægilops*) | 51 |
| MUSTARD-TREE (*Salvadora persica; Sinapis nigra*) | 57 |
| MYRTLE-TREE (*Myrtus communis*) | 62 |
| OLIVE-TREE (*Olea europæa*) | 65 |
| OIL-TREE (*Elæagnus angustifolia*) | 70 |
| PALM-TREE (*Phœnix dactylifera*) | 72 |
| POMEGRANATE-TREE (*Punica Granatum*) | 77 |
| SHITTAH-TREE (*Acacia Seyal*) | 80 |
| SYCOMORE-TREE (*Ficus sycomorus*) | 85 |
| TEIL-TREE, OR TEREBINTH-TREE (*Pistacia Terebinthus*) | 89 |
| HUSK-TREE (*Ceratonia Siliqua*) | 93 |
| PLANE-TREE (*Platanus orientalis*) | 97 |
| NUTS (*Juglans regia; Pistacia vera*) | 100 |
| VINE (*Vitis vinifera*) | 103 |
| WILLOW-TREE (*Salix babylonica*) | 108 |
| CAMPHIRE (*Lawsonia inermis*) | 113 |
| ALMUG OR ALGUM TREE (*Santalum album; Pterocarpus santalinus*) | 115 |
| ALOES-TREE, OR LIGN-ALOES TREE (*Aquilaria Agallochum*) | 116 |
| ASH-TREE | 117 |
| EBONY-TREE (*Diospyros ebenus*) | 117 |
| JUNIPER-BUSH (*Genista monosperma*) | 119 |
| POPLAR (*Populus alba*) | 120 |
| MYRRH-TREE (*Balsamodendron Myrrha*) | 121 |
| ESHEL (*Tamarix orientalis*) | 122 |
| THYINE-WOOD (*Xylon thyinum*) | 124 |
| APPLE-TREE (*Pyrus malus*) | 126 |
| THORNS AND BRIERS AND BRAMBLES | 128 |
| LOT, OR LADANUM (*Cistus creticus*) | 129 |

# CONTENTS.

| | |
|---|---|
| STACTE (*Nataf*) | 130 |
| PINE-TREE (*Tidhar*) | 131 |
| ANISE OR DILL (*Peucedanum graveolens; Anethon*) | 133 |
| BEANS (*Vicia faba; Pol; Cyamos; Faba vulgaris*) | 135 |
| SWEET CANE (*Andropogon calamus-aromaticus*) | 137 |
| CORIANDER (*Coriandrum sativum*) | 139 |
| CORN | 142 |
| CUMMIN (*Cuminum cyminum*) | 143 |
| FITCHES (*Nigella sativa*) | 146 |
| FLAX (*Linum usitatissimum*) | 149 |
| FRANKINCENSE (*Boswellia thurifera*) | 154 |
| GALBANUM (*Polylophium officinale*) | 156 |
| WILD GOURD (*Citrullus colocynthis*) | 158 |
| HEMP (*Cannabis sativa*) | 162 |
| SAFFRON (*Crocus sativus*) | 165 |
| LENTILES (*Ervum lens*) | 167 |
| RUE (*Ruta graveolens*) | 170 |
| MINT (*Mentha sylvestris*) | 173 |
| ROSE (*Narcissus tazetta*) | 175 |
| MILLET (*Panicum miliaceum*) | 178 |
| TARES (*Lolium temulentum*) | 180 |
| LILY—OLD TESTAMENT (*Nymphœa lotus*) | 183 |
| LILY—NEW TESTAMENT (*Anemone coronaria*) | 187 |
| MELON (*Cucumis melo*) | 191 |
| NETTLE (*Urtica urens*) | 194 |
| GARLIC (*Allium sativum*) | 197 |
| GRASS | 198 |
| LEEK (*Allium porrum*) | 200 |
| ONION (*Allium cepa*) | 202 |
| WHEAT (*Triticum sativum; var. compositum*) | 204 |
| SPELT (*Triticum spelta*) | 209 |
| BARLEY (*Hordeum distichon*) | 211 |
| COCKLE (*Baoshah*) | 214 |
| GOURD (*Ricinus communis; Cucurbita pepo*) | 215 |
| CUCUMBER (*Cucumis sativus*) | 219 |
| BULRUSH AND RUSH (*Papyrus antiquorum*) | 222 |
| SPIKENARD (*Nardostachys jatamansi*) | 226 |
| COTTON (*Gossypium herbaceum*) | 229 |
| REED (*Arundo donax*) | 233 |
| FLAG (*Cyperus esculentus*) | 235 |
| DOVE'S DUNG (*Ornithogalum umbellatum*) | 237 |
| MANDRAKE (*Atropa mandragora; Mandragora officinalis*) | 239 |
| THISTLE (*Tribulus terrestris*) | 242 |
| HEMLOCK | 244 |
| WORMWOOD | 246 |
| BITTER HERBS | 247 |
| CONCLUSION | 248 |

# THE PLANTS OF THE BIBLE.

## ALMOND-TREE.

(*Amygdalus communis, Linn.; Prunus amygdalus, Stokes.*)

"The almond tree shall flourish."—Eccles. xii. 5.

THE almond-tree is the *Amygdalus communis* of botanists. It is referred to in Scripture under the Hebrew name of *shaked*, apparently derived from the word *shakad*, meaning haste or waking early, with reference to its early blossoming. The Hebrew word *luz* (from the Arabic *louz*), which occurs in Genesis xxx. 37, and which has been translated "hazel," is considered to be another name for the almond. *Luz* is supposed to refer to the tree, and *shaked* to the fruit of the almond. Rosenmüller thinks that the former name designates the wild tree, and the latter the cultivated one. The tree belongs to the natural order Rosaceæ, the Rose family. It is included under the section (sub-order) Amygdaleæ or Drupiferæ of that family,—distinguished by the nature of the fruit, which has a kernel, enclosed in a shell or stone, as it is called, and surrounded by a more or less succulent covering. In this section are included also the peach, the nectarine, the apricot, the plum, and the cherry. The leaves of the tree are long and narrow, with an acute point

BRANCH OF ALMOND-TREE.

and saw-like margin. The tree is a native of Asia and Barbary. It is cultivated extensively in the south of Europe, and is also met with in gardens in Britain. It was probably not a native of Egypt; for Jacob, when sending his sons to that country, told them to take almonds (*shakedim*) as a present to Joseph (Gen. xliii. 11). The tree appears also to have continued to produce fruit during the period of famine in the land of Canaan.

The almond-tree blossoms very early in the season. Kitto mentions it among the trees of Palestine that flower in January. The flowers are of a pinkish colour,

ALMOND-TREE.—(*Amygdalus communis.*)

and are produced before the leaves, and are therefore very conspicuous. This hastening of the period of flowering seems to be alluded to in Jeremiah i. 11, 12—"What seest thou? And I said, I see a rod of an almond tree [*shaked*]. Then said the Lord unto me, Thou hast well seen: for I will hasten [*shaked*] my word to perform it."

In Ecclesiastes xii. 5, it is said, "The almond tree shall flourish." This has often been supposed to refer to the resemblance between the flowers of the almond and the hoary locks of old age. But this interpretation is not borne out by an examination of the blossom of the almond, which is pinkish, and not pure white. The passage rather appears to refer to the hastening of old age. As the almond-tree ushers in spring, so do the signs referred to in the context indicate the coming of old age and death. In both passages the tree is taken as denoting the speedy approach of marked epochs or events.

The almond-tree was associated in the minds of the children of Israel with the choosing of the house of Levi for the service of the tabernacle. Moses, we are told in Numbers xvii., laid up the twelve rods of the princes of Israel before the Lord, in the tabernacle of witness; and on the morrow the rod of Aaron, which represented the house of Levi, brought forth buds and blossoms, and yielded almonds. Thus the Lord made to cease the murmurings of the children of Israel against Moses. This rod was deposited, as a memorial, in the ark of the covenant (Heb. ix. 4).

In Northern Europe almonds are mentioned about the middle of the second century before Christ. They were produced largely in the islands of the Greek Archipelago. The Knights Templars in Cyprus levied tithes of almonds in 1411. In medieval cookery the consumption of almonds was very great.

The fruit of the almond was used to furnish a model

for certain kinds of ornamental carved-work. Thus, in speaking of the candlestick in the tabernacle, Moses says that its bowls were made like unto almonds (Exod. xxv. 33, 34; xxxvii. 19, 20). Pieces of crystal called "almonds" are still used by manufacturers in the adorning of cut-glass chandeliers. The kernel of the almond is used for food, and for supplying oil. There are two varieties of the tree, one yielding sweet and the other bitter almonds. These two varieties are very like each other, and can scarcely be distinguished at first sight. Sweet almonds (*Amygdalus communis*, variety *dulcis*) contain a fixed oil and emulsine; while bitter almonds (*Amygdalus communis*, variety *amara*) contain, in addition, a nitrogenous substance called amygdaline, which, by combination with emulsine, produces a volatile oil and prussic acid. Bitter almonds, when eaten in small quantity, sometimes produce nettle-rash, and when taken in large quantity they may cause poisoning.

Let the almond-tree be the means of calling our attention to the hasting of God's Word, so that we may be ready, having our loins girt and our lamps burning, when the Lord comes.

In the figures a representation is given of the almond-tree, and of its blossoms and fruit.

Almonds of all kinds imported into the United Kingdom are reported as follows:—

|      | Cwts.  |
|------|--------|
| 1876 | 77,196 |
| 1877 | 60,547 |
| 1878 | 58,360 |
| 1879 | 46,319 |
| 1880 | 86,763 |

# BOX-TREE.

### (*Buxus sempervirens, Linn.*)

---

"I will set in the desert the fir-tree, and the pine, and the box-tree together."—Isa. xli. 19.

THE box-tree is the *Buxus sempervirens* of botanists. According to some the Palestine plant differs from the common box in the form and size of its leaves. It is mentioned in the Bible under the Hebrew name of *teasshur*. The tree belongs to the natural order Euphorbiaceæ, the Spurgewort family. The plants of this order have peculiar involucrate flowers, often without any perianth, and their fruit is usually composed of three carpels, which separate in an elastic manner when ripe. They abound in milky juice, which has in general acrid and poisonous qualities. Starch, as well as oils and caoutchouc, are procured from many of the species.

The box is a native of most parts of Europe, and grows well in England, as at Boxhill, in Surrey. It is prized as an ornamental evergreen; and in a dwarf state is used for garden borders. Its wood, imported from the Levant, is used by the wood-engraver, the turner, the mathematical instrument maker, the comb and toy maker, and others. The wood is hard and durable, and was formerly made into tablets which were covered with wax and used for writing. The practice of inlaying

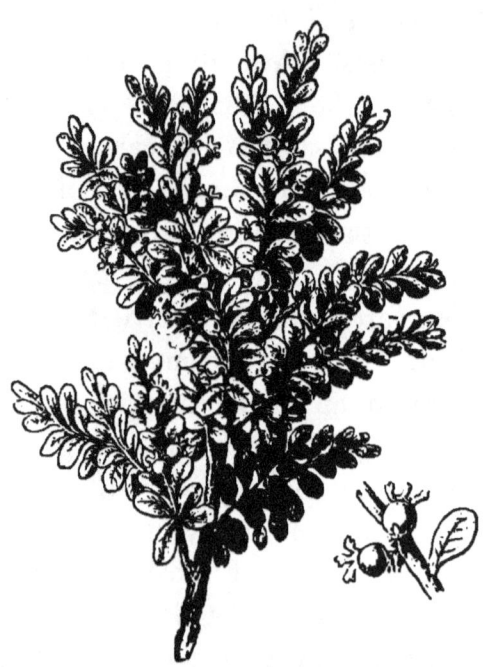

BOX-TREE.—(*Buxus sempervirens.*)

box-wood with ivory is noticed by ancient authors. Thus Virgil says :—

> "Aut collo decus, aut capiti; vel quale per artem
> Inclusum buxo, aut Oriciâ terebintho,
> Lucet ebur."—*Æneid*, x. 135.

[An ornament, either for the neck or for the head; or as shines the ivory by art enchased in boxwood or Orician ebony.]

The prophet Isaiah refers to the box as one of the

trees fitted to beautify the wilderness and the desert: "I will plant in the wilderness the cedar, the shittah tree [*Acacia Seyal*], the myrtle, and the oil tree [olive tree]; I will set in the desert the fir tree, the pine, and the box tree together" (xli. 19). Again, in referring to the glory of the latter days, he speaks of the box as adorning the Lord's temple: "The glory of Lebanon shall come unto thee, the fir tree, the pine tree, and the box together, to beautify the place of my sanctuary" (lx. 13). Royle says: "The box-tree, being a native of mountainous regions, was peculiarly adapted to the calcareous formations of Mount Lebanon, and therefore likely to be brought from thence with the coniferous woods for the building of the temple; and it was well suited to change the face of the desert." The prophet's prediction, however, seems to have reference to the trees of righteousness, the planting of the Lord (lxi. 3), and to bring before us the members of Christ's true Church, differing in many particulars, but all enjoying sweet communion, and worshipping the Lord together.

The prophet Ezekiel (xxvii. 6), when describing the commerce of Tyre, uses the word *ashur*, which, by most commentators, is supposed to be a contraction of *teasshur*, or box. The translation of the passage should probably be—" Of the oaks of Bashan have they made their oars; the benches of the rowers have they made of *ashur-wood* [box-wood], inlaid with ivory, brought out of the isles of Chittim [the isles of Greece]." Thus, in place of *Ashurites*, as in our Authorized Version, the word *ashur-wood* ought to be substituted. It is conjectured that Corsica and Sardinia may have been included among the isles of Chittim whence box-wood was brought to Judea. Pliny and Theophrastus mention that Corsica was famous for its box-trees. Another species, called *Buxus balearica*, Turkey-box, is found in the Balearic Isles. Its wood is also much used.

By some the *teasshur* has been confounded with the *sherbin*, which is in reality a species of juniper (*Juniperus phœnicea*), found on Lebanon, and sometimes erroneously called cedar. By some the Palestine-box is thought to be a small species or variety named *Buxus tenuifolia*, with more slender leaves.

# BAY-TREE.

### (*Laurus nobilis, Linn.*)

---

"Spreading like a green bay tree."—Ps. xxxvii. 35.

HE plant called bay-tree in the Bible is supposed to be the sweet-bay,—the *Laurus nobilis* of botanists. It belongs to the natural order Laurineæ, the Laurel family. The laurels are aromatic and fragrant plants, yielding fixed and volatile oils, as well as camphor. They have dotted leaves; stamens partly fertile and partly abortive, the former having anthers opening by valves; and their fruit is a berry or drupe. The sweet-bay—the *ezrach* of the Hebrew—is an evergreen tree, attaining the height of twenty or thirty feet, common in the south of Europe, and found also in Palestine. At the present day it is said to luxuriate in the old gardens of Tyre and Sidon, and beside some forgotten towers and deserted winepresses in the Holy Land.

The tree yields a green oil, denominated oil of bays. Its branches were used for crowning the victors in the ancient games of Greece and Rome, as well as for decorating the brow of the poet.

This and the other species of true laurel must not be confounded with the plants commonly called laurels in gardens. The latter consist of the cherry laurel and the Portugal laurel, which belong to a totally different order of plants—namely, to the same section of the rose family as the almond and the plum. The ratafia odour

BAY-TREE.—(*Laurus nobilis.*)

emitted by the bruised leaves of these garden laurels is very different from the aromatic perfume given out by the sweet-bay leaves. The cherry-laurel water furnished by the large-leaved bay-laurel contains prussic acid, and has consequently poisonous qualities. In this respect the plant resembles the bitter almond. Another garden plant, denominated Laurustinus, must also be distinguished from the sweet-bay; it is the *Viburnum Tinus* of botanists, and belongs to the natural order *Caprifoliaceæ.*

The psalmist, in Psalm xxxvii. 35, thus alludes to the laurel now under consideration: "I have seen the wicked in great power, and spreading himself like a green bay tree." The vigour and beauty of the tree made it a fit emblem of prosperity; and its association with the fame of the victor and the poet suggested the idea of the honour which cometh from man.

Royle says: "The cause why the laurel is not more frequently mentioned in Scripture is probably because it was never very common in Palestine; as otherwise, from its pleasing appearance, grateful shade, and the agreeable odour of its leaves, it could hardly have failed to attract attention." In the neighbourhood of Antioch the tree is said to be abundant. Hasselquist suggests that the rosebay, the *Nerium Oleander* of botanists, might be the plant referred to by the psalmist. It grows by the sides of streams in some parts of Judea, and is conspicuous alike for its foliage as for its showy flowers. The perfume of the oleanders around the Lake of Tiberias has attracted the notice of travellers. Royle and others think that the oleander is the *rhodon*, or rose, of the Apocrypha.

Some commentators suppose that the term *ezrach* applies to a tree grown in its native soil, and not to any special tree, such as the bay.

# CEDAR-TREE OF LEBANON.

(*Cedrus Libani, Linn.*)

---

"The boughs thereof were like the goodly cedars."—Ps. lxxx. 10.

HE cedar-tree of Lebanon is noticed in the Bible under the Hebrew name of *eres* or *œres*. It is probable, however, that this name was also applied to other allied plants. The Arabs call the tree *arz* or *ars*. It is the *Cedrus Libani* of botanists, and belongs to the natural order Coniferæ, the Cone-bearing family, in which it is associated with the pines, firs, spruces, and larches.

In early times, the cedar appears to have grown abundantly on Lebanon, and to have proved its distinguishing feature. Hence it was called "the glory of Lebanon" (Isa. xxxv. 2 ; lx. 13). In various passages of the Old Testament, we read of the cedars of Lebanon sent by Hiram, king of Tyre, for the building of David's house, of the temple at Jerusalem, and of Solomon's house (2 Sam. v. 11, vii. 2, 7 ; 1 Kings v. 6, 8, 10, vi. 9, 10, 15, 16, 18, 20, vii. 2, 3, 7, 11, 12, ix. 11 ; 1 Chron. xvii. 6 ; 2 Chron. ii. 8). Beams, boards, pillars, walls, floor, ceiling, throne, and altar of cedar are mentioned. This timber was employed in consequence of its superior quality. It is stated that Solomon "made cedars to be as the sycomore trees [sycomore fig-trees] that are in the vale [or in the low plains], for abundance" (1 Kings x. 27; 2 Chron. ix. 27). Travellers tell us that there are

CEDAR CONES.

still numerous cedars on Lebanon. There are at least nine distinct localities, containing many thousand trees and numerous saplings. Sir J. D. Hooker says that "cedars are found on the mountains of Algeria, on the whole range of Taurus, and in the Kedesha valley of Lebanon. In the Kedesha valley the number of trees is about four hundred. They are of various sizes, from about eighteen inches to upwards of forty feet in girth." He calculates the age of the Kedesha cedars at eight hundred years.

Robinson says: "Cedars of Lebanon, still called *arz*, stand mostly on four small contiguous rocky knolls, within a compass of less than forty rods in diameter. They form a thick forest without underbrush. The older trees have each several trunks, and thus spread themselves widely around." Some of the older trees are much broken, and will soon be destroyed.

Burckhardt, in 1810, says of this spot: "Of the oldest and best-looking trees I counted eleven or twelve, twenty-five very large ones, about fifty of middling size, and more than three hundred smaller and young ones. In 1843 Dr. Wilson counted twelve of the ancient trees not standing together, and of the younger growth three hundred and twenty-five." (*Lands of the Bible*, ii. 389.)

In 1853 Ritter numbers four hundred in all, of which twelve are spoken of as the largest.

During the last three centuries the number of the older trees has diminished by about one-half.

In 1550 Belon counted twenty-eight; in 1556 Füren counted twenty-five; in 1575 Rauwolf counted twenty-four, and two others the boughs of which were broken off by age; in 1596 Dandini counted twenty-three; in 1632 Roget counted twenty-two; in 1660 D'Arvieux counted twenty-three; in 1688 De la Roque counted twenty; in 1696 Maundrell counted sixteen; in 1738 Korte counted eighteen very old and large; in 1739 Pococke counted fifteen, and one recently blown down; in 1755 Schulz counted twenty. (*Büsching Edbeschr*, xi. i. 314.)

The cedars stand in a magnificent amphitheatre on Lebanon, about 6,400 feet above the level of the sea, with the ridges of the mountains rising two to three thousand feet above it. The amphitheatre fronts the west, and the snow extends round from south to north. It appears, according to Ehrenberg, that cedars are found on the northern parts of Lebanon also.

The cedar of Lebanon is a wide-spreading evergreen tree, from fifty to eighty feet in height, with numerous large horizontal branches. Ezekiel, when describing the cedar, speaks of its high stature, "its top among the thick boughs, its multiplied boughs, its long branches, and its shadowing shroud" (Ezek. xxxi. 3–9). The "goodly cedars," or cedars of God, are mentioned in Ps. lxxx. 10; and "excellent cedars," in Song of Sol. v. 15. Isaiah speaks of the cedars of Lebanon being "high and lifted up" (ii. 13); and of the "tall cedars" (xxxvii. 24). As the branches extended so did the roots, and thus the tree was firmly fixed in the soil, and enabled to withstand the violence of storms. Hence the prophet Hosea says, "He shall cast forth his roots as Lebanon" (xiv. 5). The watering of the roots by means of the streams of Lebanon is referred to by Ezekiel in the passage already noticed. The tree was distinguished for its exalted and vigorous growth; hence it is singled out among those of which Solomon wrote: "He spake of trees, from the cedar tree that is in Lebanon, even unto the hyssop [caper-bush] which springeth out of the wall" (1 Kings iv. 33). The righteous are represented as growing like the cedar-trees of Lebanon (Ps. xcii. 12); and Israel like the cedar-trees beside the waters (Num. xxiv. 6). The wood of the cedar is reddish white, and is easily worked. The tree yields a sweet-smelling resin, which is alluded

CEDARS OF LEBANON.

to in Scripture as "the smell of Lebanon" (Song of Sol. iv. 11; Hos. xiv. 6).

It has been supposed that the cedar-wood mentioned in Leviticus xiv. 4, and Numbers xix. 6, was the produce of a fragrant species of juniper plentiful in the desert, and growing in crevices of Sinai. The cedar-wood used for pencils at the present day is the produce of *Juniperus bermudiana*, a native of the West Indies. In some heathen countries species of juniper are used as incense on account of their fragrance. *Pinus halepensis* and *Juniperus excelsa* grow along with cedars on Lebanon.

Other cedars are *Cedrus deodara* of the Himalaya and *Cedrus atlantica* of the Atlas Mountains. By some the three cedars are supposed to belong to one species.

Cedar is also mentioned in the following passages:—
2 Kings xix. 23; Ezra iii. 7; Song of Sol. v. 17, viii. 9; Isa. ix. 10, xiv. 8, xliv. 14; Jer. xxii. 7, 14, 23; Ezek. xvii. 3, 22, 23, xxvii. 5; Amos ii. 9; Zech. xi. 1, 2.

# HEATH-TREE (SAVIN).

*(Juniperus Sabina, Linn.)*

---

"He shall be like the heath in the desert."—JER. xvii. 6.

THE Hebrew word *arar*, translated "heath," occurs in two passages in the Bible—Jeremiah xvii. 6, already quoted, and Jeremiah xlviii. 6, where it is said, "Flee, save your lives, and be like the heath in the wilderness." The best authorities refer this to a species of *juniper* which grows in the rocky parts of the desert, and is called *Juniperus Sabina*, known as savin. It belongs to the natural order Coniferæ, the Cone-bearing family, and the sub-order Cupressineæ (cypresses). It is a stunted shrub, which, like the other juniper, bears a succulent cone, and has a strong turpentine flavour. The volatile oil procured from its branches and leaves has dangerous qualities. Dr. Tristram remarks: "Its gloomy, stunted appearance, with its scale-like leaves pressed close to its gnarled stem, and cropped close by the wild goats, as it clings to the rocks about Petra, gives great force to the contrast suggested by the prophet between him that trusteth in

HEATH-TREE.—(*Juniperus Sabina*.)

man, naked and destitute, and the man that trusteth in the Lord, flourishing as a tree planted by the waters."

# CINNAMON-TREE AND CASSIA-TREE.

(*Cinnamomum zeylanicum*, Breyne; and *Cinnamomum Cassia*, Bl.)

---

"Calamus and cinnamon, with all trees of frankincense."
SONG OF SOL. iv. 14.

CINNAMON is mentioned in several places in the Old Testament, under the Hebrew name of *kinnamon*. The plant is *Cinnamomum zeylanicum* of botanists. It belongs to the natural order Laurineæ, the Laurel family. In this order are found many aromatic plants, yielding volatile oils and tonic barks. (See the description of the *Bay-tree*.) The plant grows in India; and its bark, under the name of cinnamon, is imported at the present day from Ceylon, and also from the Malabar coast, in bales and chests, the bundles weighing about one pound each. It was imported into India by the Phœnicians or the Arabians. It is distinguished from other allied species by its acuminated tricostate leaves, the ribs coming into contact at the base, but not uniting. The best cinnamon is procured from branches three years old. The outer bark is of a whitish-gray colour, and is nearly tasteless, while the inner bark constitutes the cinnamon, which is imported in a quilled form into Britain. Oil of cinnamon is obtained from the bark by distillation, after it has been macerated in sea-water; and a fatty matter is procured from the fruit by boiling. This fat was used by the Portuguese in making candles.

CINNAMON-TREE.—(*Cinnamomum zeylanicum.*)

Cinnamon was highly valued as a spice and perfume. It was one of the principal spices employed in the manufacture of precious ointment for the tabernacle (Ex. xxx. 22–25). Solomon speaks of it also as one of the frankincense plants: " Calamus and cinnamon, with all trees of frankincense" (Song iv. 14). Its use as a perfume is referred to in Prov. vii. 17: "I have perfumed my bed with myrrh, aloes [*Aquilaria Agallochum*], and cinnamon." And the merchandise of it is noticed in the account of the destruction of the Apocalyptic Babylon:

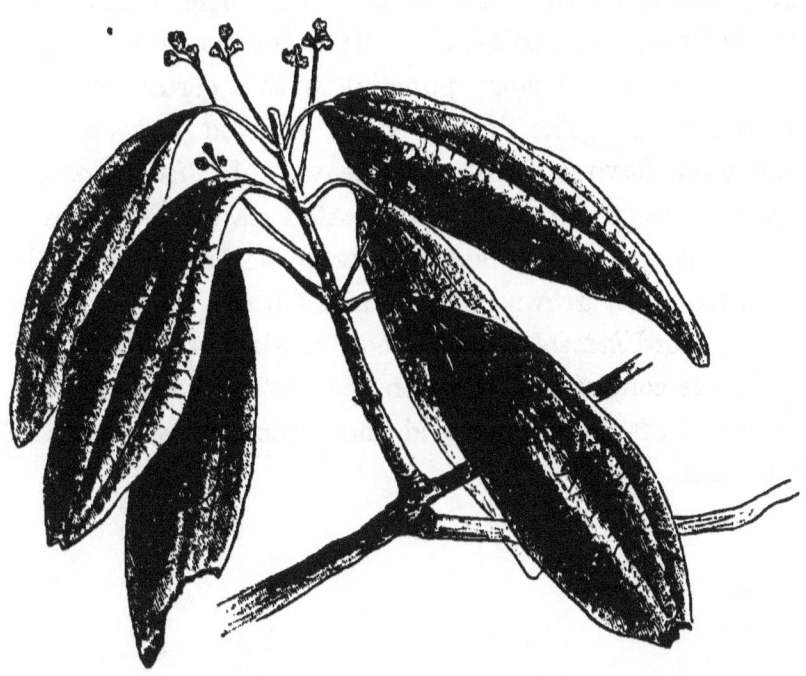

CASSIA-TREE.—(*Cinnamomum Cassia.*)

"Cinnamon, and odours, and ointments, and frankincense" (Rev. xviii. 13).

Besides the true cinnamon plant, we must also refer to another species known under the name of *cassia*. It is mentioned in Scripture as *kiddah*. It constituted one of the ingredients in the holy ointment already referred to (Ex. xxx. 24); and it is recorded by Ezekiel among the merchandise of Tyre (Ezek. xxvii. 19). The plant referred to in these passages appears to be the *Cinnamomum Cassia* of botanists, which is distinguished from the *Cinnamomum zeylanicum* by its oblong-lanceolate triplicostate leaves, the three ribs uniting together for some extent at the base of the leaf. The bark of the tree is known as cassia-bark. It is inferior to cinnamon, being coarser and more pungent, with a certain amount of bitterness. The leaves when chewed have a true cinnamon flavour, while the leaves of *Cinnamomum zeylanicum* when similarly treated taste like cloves. Cassia-oil and cassia-buds appear to be produced by the same tree. It grows in India and China.

The word *ketzioth*, translated " cassia," in Ps. xlv. 8, is by Royle conjectured to mean the *costus* of the ancients, the *koost* of the Arabs, and the *Aplotaxis auricula* of botanists.

# FIR-TREE.

*(Cupressus sempervirens, Linn.)*

---

"I am like a green fir tree."—Hos. xiv. 8.

THE fir-tree is noticed in the Bible under the Hebrew names of *berosh* and *beroth*. Most commentators believe that the tree alluded to is the cypress, *Cupressus sempervirens* of botanists. It belongs to the natural order Coniferæ, the Cone-bearing family, sub-order Cupressineæ. These coniferous trees are resinous in their nature; their leaves are very narrow and sharp-pointed (hence called needle-trees by the Germans); their staminate flowers are in deciduous catkins; and their pistillate flowers in cones, the scales of which cover one, two, or more naked seeds. The wood of the tree is marked with remarkable dots or discs, which are easily seen under the microscope. The tree has a tapering form not unlike that of the Lombardy poplar; and in southern latitudes it attains a height of fifty or sixty feet. Its fruit is a more or less rounded cone, flattened at the apex, and composed of peltate (shield-like) scales, covering numerous winged seeds. Its timber is durable. The gates of Constantinople, which stood for more than a thousand years, were made of it. The tree is a native of Greece, Asia Minor, Syria, and Palestine. The Mohammedans plant it in their burying-grounds.

Allusion is frequently made in the Bible to the vigorous

BRANCH OF CYPRESS-TREE.—(*Fir-tree of Scripture.*)

growth of the fir-tree. Thus Ezekiel, when describing the power of the Assyrian, selects the fir-tree on account of its noble growth, and says, "The fir-trees were not like his boughs" (xxxi. 8). For the same reason it is associated with the cedar of Lebanon. Sennacherib, the king of Assyria, is represented as saying, "With the multitude of my chariots I am come up to the height of the mountains, to the sides of Lebanon, and will cut down the tall cedar trees thereof, and the choice fir trees thereof" (2 Kings xix. 23 ; Isa. xxxvii. 24).

The wood was used for various purposes, such as in house-building, ship-building, the formation of musical instruments, etc. It was one of the kinds of timber sent by Hiram to Solomon for the construction of the temple (1 Kings v. 8, 10, ix. 11 ; 2 Chron. ii. 8). The floor of the house was covered with planks of fir; and the two doors at the entrance of the temple, and the ceiling, were made of the same kind of wood (1 Kings vi. 15, 34; 2 Chron. iii. 5). Rafters of *berosh* are also referred to (Song i. 17). David and all the house of Israel played on musical instruments made of *berosh* wood.

Fir-trees are mentioned in connection with the future renovated earth. Isaiah says, " I will set in the desert the fir tree" (xli. 19) ; " Instead of the thorn shall come up the fir tree" (lv. 13); " The glory of Lebanon shall come unto thee, the fir tree, the pine tree, and the box tree together, to beautify the place of my sanctuary" (lx. 13).

The word *tirzah*, translated " cypress" in Isa. xliv. 14, is supposed by many to mean the evergreen oak, *Quercus Ilex*, the wood of which was constantly employed by the ancients in making images. (See *Oak-tree*.)

The *gopher-wood* of which the ark was constructed (Gen. vi. 14) is supposed to be the produce of the cypress, or of some other tree belonging to the pine tribe.

Along with the cedars on Lebanon there is found a fir-tree called *Pinus halepensis* (Aleppo pine), and it may

CYPRESS-TREE (FIR-TREE OF SCRIPTURE).

be referred to in some passages in the Bible. It is common in Western Palestine, and is one of the characteristic trees of Lower Lebanon. It is said to be espe-

ALEPPO PINE.

cially the fir-tree of Palestine, and is only inferior to the cedar in size. Another fir, *Pinus maritima*, is found on the sandy plains near the coast, and helps to keep the loose sand from being blown over the country.

# FIG-TREE.

(*Ficus Carica, Linn.*)

---

"Learn a parable of the fig tree."—MATT. xxiv. 32.
"Now from the fig tree learn her parable."—*New Translation.*

THE Hebrew word *teenah*, and the Greek word *sycé* or *sucé*, are translated "fig" and "fig-tree" in Scripture. The tree is called by botanists *Ficus Carica*. It belongs to the natural order Artocarpaceæ, the Bread-fruit family, and the sub-order Moreæ, which includes also the mulberry. The tree is characterized by its fruit, which is formed by an enlarged succulent hollow receptacle, containing the flowers in its interior. Hence the flowers of the fig-tree are not visible until the receptacle is cut open. The tree is a native of the East, and has been transported into Europe. It is grown in the south of Europe, including Greece and Italy; and in Northern and Western Africa. A wild type is known in Italy by the name of *Caprifico*.

Figs have been cultivated from the earliest times. The fig is the first tree mentioned by name in Scripture

(Gen. iii. 7). The figs of Athens were celebrated for their flavour. Figs at the present day are brought to this country from Smyrna in small boxes called *drums;* the quantity imported in 1858 was nearly seventeen hundred tons.

The fig-tree was common in Palestine, which was

BRANCH OF FIG-TREE.

described as being "a land of wheat, and barley, and vines, and fig trees, and pomegranates" (Deut. viii. 8). The parties who went from the wilderness of Paran to search the land "brought of the pomegranates and of the figs" (Num. xiii. 23). The fig-tree is employed to indicate the peace and prosperity of a nation, 1 Kings

iv. 25; also Micah iv. 4, where it is said, "They shall sit every man under his vine and under his fig-tree." Sennacherib, king of Assyria, employs the same metaphor in order to induce the inhabitants of Jerusalem to surrender (2 Kings xviii. 31; Isa. xxxvi. 16). No tree furnishes better protection from the rays of the sun in Eastern countries than the fig-tree. Figs constitute an important article of food in Eastern countries, and are eaten both in a fresh and in a dried condition. In the latter state they are spoken of as being made into cakes, called *debelim*. Abigail brought two hundred cakes of figs to David and his men (1 Sam. xxv. 18); and the armies that came to David in Hebron brought cakes of figs (1 Chron. xii. 40). A piece of a cake of figs was given to the Egyptian who was found in a famishing state in the field (1 Sam. xxx. 12). Good and bad figs are used by Jeremiah as emblems of good and evil (Jer. xxiv.). Tristram mentions that in the island of Cyprus there are clumps of fig-trees round each door, and that he enjoyed rest and food beneath the shade of the Cyprian fig-tree. Under such a fig-tree, he remarks, Nathanael had wrestled in prayer, and was convinced at once of the Messiahship of Jesus by His knowledge of his retirement (John i. 48, 49).

The failure, destruction, and falling of the figs are mentioned as indications of the judgments of the Lord (Ps. cv. 33; Isa. xxxiv. 4; Jer. v. 17, viii. 13; Hosea ii. 12; Joel i. 7, 12; see also Rev. vi. 13). Figs were used as a laxative, and also as a poultice. Thus Isaiah ordered a lump of figs to be laid on the boil with which

Hezekiah was afflicted, and he recovered (2 Kings xx. 7; Isa. xxxviii. 21).

Different crops of figs are produced during the year. Early figs appeared in spring—before the leaves expanded (Jer. xxiv. 2). Isaiah, Hosea, and Nahum refer to the early or first ripe figs [*bikhurah*], or the hasty fruits before the summer (Isa. xxviii. 4; Hosea ix. 10; Nahum iii. 12). "When his branch is yet tender, and putteth forth leaves, ye know that summer is nigh" (Matt. xxiv. 32). The early green fruit is alluded to in the Song of Solomon, ii. 13. Besides the forward figs of spring, there were also summer and autumn figs. When Jesus was proceeding from Bethany to Jerusalem, "he hungered. And when he saw a fig tree in the way, he came to it, and found nothing thereon, but leaves only" (Matt. xxi. 18, 19). The period was early (end of March or beginning of April), and, according to Mark, "the time of figs was not yet" (Mark xi. 13); still, as the tree was in full leaf, it might have been expected that some early figs would have been found. Finding no appearance whatever of fruit, however, our Saviour said to the tree, "Let no fruit grow on thee henceforward for ever. And presently the fig tree withered away."

# HYSSOP.

*(Capparis spinosa, Linn.; Caper-plant and its variety, Capparis ægyptiaca, Lam.)*

---

"He spake of trees, from the cedar tree that is in Lebanon, even unto the hyssop that springeth out of the wall."—1 KINGS iv. 33.

THE Hebrew word *esobh* or *ezob* and the Greek *hyssopos* are translated "hyssop" in the Bible. There have been great differences of opinion regarding the nature of the plant thus mentioned by the sacred writers of the Old and New Testaments. Some have thought that it was a minute moss or fern, or some other wall-plant; others, that it was the plant called hyssop at the present day, or one allied to it, such as rosemary, marjoram, or thyme. Some authors (as Royle) think that the word *abizonah* (translated "desire"), which only occurs in Ecclesiastes xii. 5, is the caper-plant. The passage is as follows: "When the almond tree shall flourish, and the grasshopper shall be a burden, and *desire* [abizonah] shall fail: because man goeth to his long home." The rabbins apply the term *abunott* to the small fruit of trees and to berries as well as to that of the caper-bush, which is common in Syria and Arabia. The fruit was apparently calculated to excite *desire*. Celsius, however, in his "Hierobotanicum," does not agree with this interpretation of "desire." He remarks that Solomon never speaks of capers, but of wine and perfumes. After a careful examination, Dr. Royle has come to the con-

BRANCH OF HYSSOP.

clusion that the hyssop of the Bible is the caper-plant (*Capparis spinosa* of botanists); that the name of the plant in Arabic, *azaf*, corresponds with the Hebrew *esobh*; and that the shrub is fitted for all the purposes mentioned in the Scriptures.

The caper-bush belongs to the natural order Capparidaceæ, or the Caper family. The plants of this order have pungent, stimulant, and antiscorbutic qualities. The caper-bush grows in Lower Egypt, in the deserts of Sinai, and in Palestine. The localities in which the plant delights are barren soils, rocky precipices, and the sides of walls. The caper-plant grows on walls in many southern countries. I have gathered it on walls in Italy. Tristram saw it hanging from the walls of Jerusalem, also in steep rocks in the gorge of the Kidron. He also says that the variety called *ægyptiaca* has a trailing habit on the sandy plain between Jericho and Jordan, as well as at the south-east end of the Dead Sea, and on the plains of Shittim.

Hyssop is mentioned in several passages of the Old Testament in connection with cleansing and purification. The first mention of it is in Exodus xii. 22, at the institution of the passover, where it is directed that the blood of the lamb shall be sprinkled by means of hyssop on the dwellings of the Israelites. In the cleansing of the leper and of the house affected with the plague of leprosy, hyssop was also employed in a similar way (Lev. xiv. 4–7, 49–52). It was also used in the burning of the heifer from the ashes of which the water of separation was prepared, as well as in the sprinkling of the water (Num. xix. 6). It seems to be in allusion to this sprinkling that the psalmist says, "Purge me with hyssop, and I shall be clean" (Ps. li. 7). Royle, however, thinks that David here refers to the detergent quality of the flower-buds of the plant, which constitute the capers of commerce, and are supposed to have cleansing properties.

Reference is made to hyssop in the New Testament also. Thus St. Paul alludes to the use of it in purification (Heb. ix. 19–21). The evangelist John, in the account which he gives of the crucifixion of our Lord, says, "Now there was set a vessel full of vinegar: and they filled a spunge with vinegar, and put it upon hyssop, and put it to his mouth" (John xix. 29). Here we have vinegar mentioned along with hyssop, probably as being the material used in the preparation of capers. It is obvious, also, from this passage, that the hyssop must have been a plant capable of furnishing a rod of moderate length, so that the sponge might be raised to the Saviour's lips. Such a statement, then, seems to exclude all those translations which would make the hyssop a minute plant or a small herb. The caper-bush would suit the purpose, as a stick of three or four feet long could be obtained from it. In the parallel passages of the Gospels according to Matthew and Mark, it is said that the sponge was put on a reed (Matt. xxvii. 48; Mark xv. 36), and the word hyssop is not introduced. This may be explained either by supposing that the word *kalamos*, translated "reed," was a *stick* of hyssop, or that part of a hyssop-bush was fastened upon the end of a reed or stick, and the sponge placed on it.

# ASPEN, OR TREMBLING POPLAR.

*(Populus tremula, Linn.)*

---

"The sound of a going in the tops of the mulberry trees."—2 SAM. v. 24.

THE Hebrew word *becaim* has been translated "mulberry trees." It is the plural of the word *baca*, which occurs in Psalm lxxxiv. 6. It is supposed by able commentators that the trees noticed under these names were poplars, several species of which occur in the Holy Land. Kitto says: "We know that the black poplar, the aspen, and the Lombardy poplar grew in Palestine. The aspen, whose long and flat leaf-stalks cause the leaves to tremble with every breath of wind, unites with the willow and oak in overshadowing the water-courses of Lower Lebanon, and with the oleander and acacia in adorning the ravines of Southern Palestine. The Lombardy poplar is described as growing with the walnut-trees and weeping-willow under the deep torrents of the Upper Lebanon." The Arabic word *bak*, which means "poplar," is very similar to the Hebrew *baca*.

The aspen (*Populus tremula* of botanists) is supposed

## ASPEN, OR TREMBLING POPLAR.

to be the tree indicated by the Hebrew words we have noticed. The quaking of its leaves has given origin to the name "trembling poplar" which is applied to it. The moving of the leaves seems to be referred to in the following passage: "And the Philistines came up yet again, and spread themselves in the valley of Rephaim.

ASPEN, OR TREMBLING POPLAR.—(*Mulberry-tree of Scripture.*)

And when David enquired of the Lord, he said, Thou shalt not go up, but fetch a compass behind them, and come upon them over against the mulberry trees [becaim]. And let it be, when thou hearest the sound of a going in the tops of the mulberry trees, that then thou shalt bestir thyself" (2 Sam. v. 23, 24; 1 Chron. xiv. 14, 15).

The poplar gave the name to the valley of Baca, which

is sometimes called the Valley of Weeping. Here the tree was associated with the willow and other plants which delight in a moist soil: "Who passing through the valley of Baca make it a well; the rain also filleth the pools" (Ps. lxxxiv. 6). In this shady valley the traveller to Zion was refreshed by the wells and pools of water.

The aspen belongs to the natural order Salicineæ, the Willow family. The plants of the order have their flowers in catkins, and their seeds covered with silky hairs. The trembling of the aspen leaf in the slightest breeze seems to depend on the flattening of the petiole or leaf-stalk in a vertical direction. The tree extends to northern countries, and is found in the alpine districts of Scotland. The sycamine-tree is the black mulberry. (See *Sycamine*.)

# OAK-TREE.

### (*Quercus Ægilops, Linn.*)

---

"And he [the Amorite] was strong as the oaks."—AMOS ii. 9.

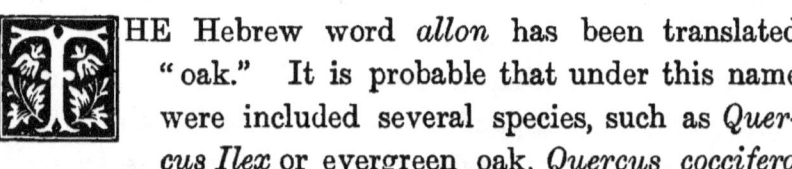HE Hebrew word *allon* has been translated "oak." It is probable that under this name were included several species, such as *Quercus Ilex* or evergreen oak, *Quercus coccifera* or kermes oak, *Quercus pseudo-coccifera* or prickly evergreen oak, *Quercus infectoria* or dyer's oak. A very abundant oak in Palestine is *Quercus pseudo-coccifera*. It covers the rocky hills with a dense brushwood eight to twelve feet high. It abounds on Mount Carmel and on the west flank of Anti-Lebanon, and in many of the valleys and slopes of Lebanon. The so-called Abraham's oak, near Hebron, is the species. Tristram says that this has for several centuries taken the place of the once renowned terebinth or teil-tree, which marked the site of Mamre on the other side of the city. The terebinth existed at Mamre in the time of Vespasian; and under it the captive Jews were sold for slaves. It disappeared about A.D. 330. The Abraham oak is the

finest tree in Farther Palestine, being twenty-three feet in girth. *Quercus Ægilops*, or valonia oak, the great prickly-cupped oak, is another species, which we have figured. It is a handsome tree, common in the Levant. It is found on Carmel and Tabor, and is the true oak of Bashan. It belongs to the natural order Corylaceæ or

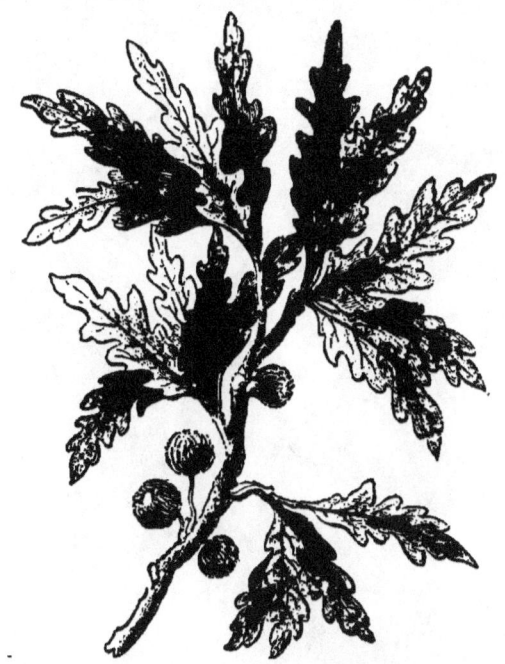

OAK-TREE.—(*Quercus Ægilops.*)

Cupuliferæ, the Hazel and Oak family. The plants of this order have their flowers in catkins, and their fruit is a nut having a cup-like covering, as in the acorn; or a husk-like covering, as in the hazel nut. The cups of the *Quercus Ægilops* are used by dyers under the name of valonia. Valonia is largely imported into Britain.

Another Hebrew word, *elah* or *ailah*, has also been

ABRAHAM'S OAK.

translated "oak" in the Bible, but it is more properly considered as meaning the terebinth-tree. Our translators have also rendered other Hebrew words by the name oak. The word translated "plain," for instance, in some passages, means an oak-grove. Thus, in 1 Samuel x. 3, in place of "the plain of Tabor," the translation ought to be, a grove of oaks at Tabor. Also in Judges ix. 37, instead of "plain of Meonenim," we should read an oak or oak-grove of the magicians. Other texts in which "plain" occurs in place of "oak" are—Gen. xii. 6, xiii. 18, xiv. 13, xviii. 1; Deut. xi. 30; Judges iv. 11, ix. 6.

In some parts of Palestine, oaks must have occupied a conspicuous place in the landscape. We read of the oaks of Bashan as being famous for strength, beauty, and utility. When the children of Israel departed from the Lord, they appear to have performed idolatrous rites in oak-groves. Thus we read in Hosea iv. 13 of the burning of incense upon the hills and under oaks. Isaiah (xliv. 14) speaks of the people taking the oak to make a god. When the Lord threatens judgment upon the nations, he refers often in a special manner to the oaks: "The day of the Lord shall be upon all the oaks of Bashan" (Isa. ii. 12, 13); "Howl, O ye oaks of Bashan" (Zech. xi. 2). Porter, in his interesting travels, when speaking of the mountains of Bashan, says, "Bleak and rocky at the base, they soon assume bolder outlines and exhibit grander features. Ravines cut deeply into their sides; bare cliffs shoot out from tangled jungles of dwarf ilex (oak), woven together with brambles and creeping plants; pointed cones of basalt, strewn here and there with cinders and ashes, tower up until a wreath of snow is wound round their heads; straggling trees of the great oaks of Bashan dot thinly the lower declivities, high up little groves of them appear, and higher still, around the loftiest peaks, are dense forests."

Solemn covenants were made under an oak. Joshua, when he solemnly charged the people, and announced to them the law of God, put up a stone of witness under an oak (Josh. xxiv. 26). In old times persons were sometimes buried under the shade of an oak. Thus it is stated in regard to Deborah, Rebekah's nurse, that she was buried under an oak in Beth-el, and the name of it was called Allon-bachuth, or the oak of weeping (Gen. xxxv. 8). The strength of the oak is referred to by Amos in speaking of the Amorite (Amos ii. 9). In the maritime city of Tyre, in its days of prosperity, the oak was used for making oars (Ezek. xxvii. 6).

# MUSTARD-TREE.

*(Salvadora persica, Linn., according to some; Sinapis nigra, Linn., according to others.)*

---

"A grain of mustard seed...when it is grown...becometh a tree."
MATT. xiii. 31, 32.

THE word *sinapi* is met with in the Gospels according to Matthew, Mark, and Luke, and it has been translated "mustard-tree." Much difference of opinion has existed as to the plant here intended. It is thought that it cannot be the common mustard of this country, which is an herb of annual growth; whereas the evangelists speak of the plant as a tree having branches in which the fowls of the air lodged. Thus, in Matt. xiii. 31, 32 it is said, "The kingdom of heaven is like to a grain of mustard seed, which a man took and sowed in his field; which indeed is the least of all seeds: but when it is grown it is the greatest among herbs, and becometh a tree, so that the birds of the air come and lodge in the branches thereof." Again, Mark describes it as a tree "shooting out great branches; so that the fowls of the air lodge under the shadow of it" (Mark iv. 31); and Luke says,

"The kingdom of God is like a grain of mustard seed, which a man took and cast into his garden; and it grew, and waxed a great tree; and the birds of the air lodged in the branches of it" (Luke xiii. 19). Our Lord also alludes to the smallness of the seed in Matt. xvii. 20, and Luke xvii. 6. The mustard-plant, then, was a branching tree with a small seed. Dr. Royle has examined this subject with his usual care and acuteness, and finds that the mustard-plant of Palestine at the present day is a tree which answers in every respect to the description of the sacred writers. The tree grows near Jerusalem, and most abundantly on the banks of the Jordan, and round the Sea of Tiberias. The seed is called *chardal* or *khardal*, which is the Arabic name for mustard. It is known to botanists as *Salvadora persica*, and belongs to the natural order Salvadoraceæ, which is considered as being nearly allied to the Olive family. It is found in Persia, Arabia, Palestine, and North Africa. An Indian species, *Sinapis indica* or *Koenigii*, has similar qualities, and receives the name of *kharjal*. The black mustard (*Sinapis nigra*) belongs to the natural order Cruciferæ.

The trunk of the salvadora is sometimes twenty-five feet high, with a diameter of one foot. Its branches are very numerous, spreading, and with their extremities pendulous, like the weeping-willow. The flowers are minute. The berry is very small—much less than a grain of black pepper—smooth, and red. Each fruit contains one seed, which is pungent, and is used as mustard. The fruit has an aromatic smell, and tastes like garden-cress. The bark of the root is acrid, and is used in India for causing blisters.

Some, however, still think that the black mustard-plant (*Sinapis nigra*) is referred to in Scripture, inasmuch as *Salvadora persica* is a sub-tropical plant found in the valley of En-gedi, and not a common plant in

MUSTARD-TREE.

Palestine. Tristram differs from Royle, and considers the common black mustard to be the plant referred to in the New Testament. He says that the *Salvadora persica* does not enter into Palestine. It is comparatively a foreign tree, and is confined to the north-east end of the Dead Sea. Travellers have noticed the great height of the common black mustard on the banks of the Jordan; and Dr. Thomson, in the plain of Acre, has seen it as tall as a horse and its rider.

In the New Testament the term "tree" is applied as well as the term "herb." The former may merely imply

WILD MUSTARD.

that it becomes taller than ordinary herbs. Flocks of birds frequent the mustard-plant for the sake of the seeds. The parable illustrates the increase of Christ's kingdom, which from small beginnings is destined finally to extend over the whole Earth.

Professor Hackett tells us that when crossing the plain of Akka, in Palestine, he saw before him a little grove of trees. On coming nearer they proved to be a grove of mustard-plants. Some of the trees were full nine feet high, with a trunk two or three inches in circumference, throwing out branches on every side. He wondered whether they were strong enough for the birds to "lodge in the branches thereof." Just then a bird stopped in its flight through the air, alighted on one of the limbs, which hardly moved beneath the weight, and began to warble forth a strain of sweetest music. The professor was delighted with the incident. His "doubts were charmed away;" the "least of all seeds" had actually grown into a substantial tree.

# MYRTLE-TREE.

(*Myrtus communis, Linn.*)

---

"Instead of the brier shall come up the myrtle tree."—Isa. lv. 13.

HE Hebrew word *hadas*, translated "myrtle," occurs in a few passages in the Old Testament. Royle says that the berries of the myrtle are at the present day sold in bazaars in India under the name of *hadas*. It is the *Myrtus communis* of botanists; and belongs to the natural order Myrtaceæ, the Myrtle family. The common myrtle is the most northern species of the order. It seems to have been in high repute in Eastern countries on account of its beautiful snow-white flowers, its dark-green foliage, and its pleasant odour. Its buds and berries have been used as spices, and a fragrant distilled water is prepared from its flowers. The bark and root are used for tanning Russian and Turkish leather, to which they communicate a peculiar odour. The leaves are also used to dress skins. They contain much oil, and have a dotted appearance when seen by transmitted light. Within the margin of each leaf there is a small vein

running from the base to the apex. The myrtle grew abundantly in Palestine and Syria, and it is noticed by Nehemiah as one of the trees which supplied branches for the construction of booths at the Feast of Tabernacles: "Go forth unto the mount [of Olives], and fetch olive branches, and pine branches [oil-tree branches], and myrtle branches, and palm branches, and branches of thick trees to make booths, as it is written" (Neh. viii. 15). The Hebrew word *etz'aboth*, translated "thick trees" (Lev. xxiii. 40; Ezek. xx. 28), is supposed by rabbinical writers to refer to the myrtle. In the passage already quoted from Nehemiah, both myrtle branches and thick trees are spoken of, so that there is some doubt as to the rabbinical view. Zechariah in his vision speaks of the angel of the Lord standing among the myrtle-trees, implying that they were well known and common in the country (Zech. i. 8, 10, 11). The myrtle is not a native of Britain, although it is generally cultivated in green-houses.

In this country it rarely becomes a tree, and does not blossom freely. In the north of Europe it is frequent. At the present day it occurs on the hills around Jerusalem, and in the valley of Lebanon, and it sometimes forms extensive thickets. Harris mentions myrtles growing in the valleys to the height of ten feet, and emitting an exquisite perfume. The tree sometimes attains a height of twenty feet. Horace speaks of myrtle crowns, and mentions the myrtle as a garden-plant; and Virgil states that the odour of Corydon's garden arose from the laurel and myrtle that were planted together (Ecl. ii. 54). Milton, describing the bower of Paradise, says,—

> "The roof
> Of thickest covert was inwoven shade,
> Laurel and myrtle, and what higher grow
> Of firm and fragrant leaf."

The tree is used by the prophets to indicate a change on the face of the earth, when "the knowledge of the Lord shall cover the earth as the waters cover the sea." Thus Isaiah, when speaking of that blessed epoch, says, "Instead of the thorn shall come up the fir tree, and instead of the brier shall come up the myrtle tree; and it shall be to the Lord for a name, for an everlasting sign that shall not be cut off" (Isa. lv. 13). Again, the Lord says by the prophet, "I will plant in the wilderness the cedar, the shittah tree, and the myrtle, and the oil tree" (Isa. xli. 19).

It has been stated that Hadassah, the original name of Esther, is derived from the word *hadas*, meaning myrtle. It has also been conjectured that Esther is formed from the word *as*, an Arabic name for myrtle, and *tur*, meaning fresh. The Jews employed the myrtle as an emblem of justice. They still use the myrtle in their synagogues at the Feast of Tabernacles. The dried flowers, leaves, and succulent part of the myrtle are sold in the bazaars at Jerusalem and Damascus.

# OLIVE-TREE.

*(Olea europæa, Linn.)*

---

"His beauty shall be as the olive tree."—Hos. xiv. 6.

THE olive-tree and olives are mentioned between thirty and forty times, oil-olive four times, and oil eighty times, in the Old and New Testaments. It is one of the earliest of the plants noticed in the Bible. In Genesis viii. 11 the dove is described as bringing the olive-branch to Noah: "Lo, in her mouth was an olive leaf pluckt off: so Noah knew that the waters were abated from off the earth." The olive-tree occurs in the first parable recorded in history (Judges ix. 9). Being associated with the assuaging of the waters of the Flood, the olive-branch is used as an emblem of peace and reconciliation. Oil running down from the head to the skirt of the garment is an emblem of the anointing of the Holy Spirit. The children of a good man are spoken of as seated as olive branches around his table. It can remain a long time under water without being injured. An experiment of this kind was made in the Botanic Garden of Edinburgh.

The name of the tree in Hebrew is *zait* or *sait*, or in Greek *elaia*. It is the *Olea europæa* of botanists, and belongs to the natural order Oleaceæ, the Olive family. The plants of this order have four divisions of their corolla, usually two stamens, a two-celled and two-seeded ovary, and a fleshy or dry fruit, which is often by abor-

OLIVE-TREE.—(*Olea europæa.*)

tion one-seeded. Tristram remarks that " to our Western eyes the olive is scarcely a beautiful tree; but to the Oriental the coolness of the pale-blue foliage, its evergreen freshness, spread like a silver sea along the slopes of the hills, speaks of peace and plenty, food and gladness." The olive-tree produces a large quantity of blossom. "He shall cast off his flower as the olive" (Job

xv. 33). Olive-oil is an important article of produce in Syria. The tree is common throughout Syria, but chiefly on the plains of Safet, Nazareth, and Nablous. Harvest usually takes place in September and October, the fruit being knocked off the trees with sticks. The oil is now exported in large quantity from Syria. The olive-tree has a drupaceous fruit, which was gathered for the purpose of furnishing oil, and seems to have been shaken off by beating the branches; hence in Deut. xxiv. 20 it is said, "When thou beatest thine olive tree, thou shalt not go over the boughs again; it shall be for the stranger, for the fatherless, and for the widow." Isaiah also alludes to the shaking of the olive-tree and the fruit left (Isa. xvii. 6). The outer, fleshy part of the fruit, yields the oil under pressure. A tree will yield ten to fifteen gallons. The finest oil at the present day is imported from Florence and Provence. In 1879 there were imported into Britain 36,198 tuns of olive-oil. The fruit of the olive is one of the first necessaries of life in the East.

The olive-tree is common in the south of Europe, and it abounded in the Holy Land, which was hence called a land of olive-trees, of olive-yards, and of olive-oil (Ex. xxiii. 11; Deut. vi. 11, viii. 8, xxviii. 40; Josh. xxiv. 13). Solomon gave to the servants of Hiram twenty thousand baths of oil (2 Chron. ii. 10). The Mount of Olives, so called from the abundance of these trees, is often referred to as the spot to which our Saviour retired alone or with his disciples (Matt. xxi. 1, xxiv. 3, xxvi. 30; Mark xiii. 3, xiv. 26; Luke xix. 29,

xxi. 37, xxii. 39; John viii. 1); and it was from the Mount Olivet that the disciples witnessed the ascension of their Master (Acts i. 12). In the prophecies regarding the glorious latter days, allusion is also made to the Mount of Olives (Zech. xiv. 4). Some very old olive-trees still exist on the mount. In the garden of Gethsemane some are said to have existed from the time of our Lord. The tree is of slow growth, and seldom attains a greater height than twenty or thirty feet.

There are two varieties of olive-trees, distinguished as the long-leaved, which is cultivated in the south of France and Italy, and the broad-leaved in Spain. The wild olive, called by the Greeks *agri-elaia*, was a low spiny tree, the branches of which were grafted on the cultivated olive. Hence the allusion by St. Paul in Romans xi. 17, 24. In this case the Gentiles are represented as the wild olive, which is grafted, contrary to nature, into the good olive, and thus bring forth fruit unto eternal life. The evergreen nature of the tree causes the psalmist to exclaim, " I am like a green olive tree in the house of God " (Ps. lii. 8); and Jeremiah says, " The Lord called thy name, A green olive tree " (Jer. xi. 16). The timber of the tree was used for furniture, and for ornamental carvings. The wood of the tree is beautifully grained, and it is still used for fine cabinet-work. In the temple it was used in the carvings, in forming the posts of the doors, and in the construction of the cherubim (1 Kings vi. 23, 31, 32). Its branches were employed at the Feast of Tabernacles (Neh. viii. 15). The bark of the tree has tonic properties. The oil expressed from the fruit was used in the temple and for anointing (Ex. xxv. 6, xxx. 23–25, xxxv. 14, xxxix. 37; Lev. viii. 12). The treading of the olive, and the expressing of its oil and the collecting of it in vats, are alluded to by Micah and Joel (Micah vi. 15; Joel ii. 24, iii. 13). The fatness of the olive-tree is noticed in Judges ix. 9, and in Romans xi. 17. The value of the trees required that there should be overseers to attend to them (1 Chron. xxvii. 28).

Zechariah in vision saw the two anointed ones represented as two olive-trees (Zech. iv. 11–14); and in Revelation xi. 4, the two witnesses are represented as " two olive trees...standing before the God of the earth."

The following are emblematical uses of the olive-tree :—

1. Emblem of peace and reconciliation—Noah in the ark; Christ on the Mount of Olives.

2. Emblem of beauty in its flowering—the beauty of holiness.

3. Emblem of fructification—bringing forth fruit to the praise and glory of God.

4. Emblem of the anointing of the Holy Spirit—oil running down from the head to the skirt of the garment. God shall anoint with the oil of gladness.

# OIL-TREE.

*(Elæagnus angustifolia, Linn.)*

"I will plant in the wilderness...the oil tree."—ISA. xli. 19.

HE Hebrew word *etz'shamen*, or oil-tree, occurs in three passages of the Bible. In the passage quoted above from Isaiah it is translated "oil-tree;" in Nehemiah viii. 15, it is translated "pine-branches;" and in 1 Kings vi. 23, "olive-tree." In the passage quoted from Nehemiah he tells the people to "go forth unto the mount, and fetch olive branches, and pine branches" (oil-tree branches). There is some difficulty in identifying the tree. From the best authorities it appears to be the *Elæagnus angustifolia*, the oleaster belonging to the natural order Elæagnaceæ, the plants of which order are marked by their scaly leaves.

The plant is abundant in Palestine, and yields a kind of oil which is much inferior to that of the olive. The lower sides of the leaves have a silvery appearance from the presence of scales. The flowers are minute, and the fruit is a green, bitter berry.

OIL-TREE.—(*Elæagnus angustifolia.*)

Some authors consider the oil-plant of Scripture the *Balanites ægyptiaca*, and it is figured in Smith's "Dictionary of the Bible."

# PALM-TREE.

*(Phœnix dactylifera, Linn.)*

---

"The righteous shall flourish like the palm tree."—Ps. xcii. 12.

HE palm-tree is called in Hebrew *tamar*, and in Greek *phœnix*. The date-palm is called *Phœnix dactylifera*. It is a diœcious tree, and belongs to the natural order Palmaceæ. The plants of this interesting family are characterized by their tall, usually unbranched stems; their pinnate or fan-shaped leaves; their flowers growing on a single or branched spadix, covered by a spathe; their fruit being a nut, drupe, or berry; and their seeds containing cartilaginous or hard albumen, with a small embryo in a cavity remote from the hilum. It has been calculated that some spathes contain two hundred thousand flowers. Palms are valuable plants, and furnish to man most important products, such as starch, sugar, oil, wax, fruit, coverings for habitations, materials for manuscripts, etc.

Date-palms were common in Palestine; and some cities were famous for the abundance of them. The name Phœnicia was given to the country by the Greeks and Romans, indicating "the land of palms." A town in Crete was called Phœnix (Acts xxvii. 12). The capture of Jerusalem by Titus was commemorated by Vespasian on his coins by the representation of a woman sitting dis-

DATE PALMS.

consolate under a palm-tree, and marked *Judæa capta*. Jericho was called "the city of palm-trees" (Deut. xxxiv. 3; Judges i. 16, iii. 13; 2 Chron. xxviii. 15). These palm-trees are referred to by Josephus, Strabo, Horace, and Pliny. The name Tamar is applied to a city in Palestine, near the south-west end of the Dead Sea, probably from the palm-trees near it (Ezek. xlvii. 19; xlviii. 28). Perhaps from the beauty of the palm the name Tamar was used as a woman's name (Gen. xxxviii. 6; 2 Sam. xiii. 1, xiv. 27). Some say that Tamar was Tadmor in the wilderness (2 Chron. viii. 4), afterwards called Palmyra. Hazezon-tamar and Baal-tamar are also mentioned (Gen. xiv. 7; Judges xx. 33). The former is the well-known En-gedi, on the western shore of the Dead Sea, long celebrated for its palm-groves, and mentioned by Josephus and Pliny. It is the present Ain Jidy, where there are no longer any palm-trees, although palm stems and leaves are found incrusted with carbonate of lime. Baal-tamar was near Gibeah in Benjamin. Bethany, on the eastern side of the Mount of Olives, means "the house of dates." The tree extends along the course of the Euphrates and Tigris, across to Palmyra and to the Syrian coast of the Mediterranean. It grows also in the northern parts of Africa. When growing in the desert, it indicates the presence of water. The Israelites in their journey "came to Elim, where were twelve wells of water, and three-score and ten palm trees, and they encamped by the waters" (Ex. xv. 27). In Numbers xxxiii. 9, these wells are called fountains. Tristram says that the station has generally been identified as the Wady Ghurûndel, where there are palm-trees and springs. Palm-trees are now scarce in Palestine. Stanley says that the palm breaks the uniformity of the Syrian landscape by the rarity of its occurrence: "Two or three in the gardens of Jerusalem, some few at Nablûs, one or two in the plain of Esdraelon, comprise nearly all the instances of the palm in Central Palestine." Tristram says: " In the gardens of Jenin (En-gannim, the 'garden-house,' 2 Kings ix. 27), at Nablous (Shechem), at Beisan (Bethshean), and at several sheltered villages near Nazareth, the palm still exists."

The stem of the date-palm exhibits what is called the endogenous mode of growth, the hardest part being on the outside. The leaves are pinnate, and are sometimes called branches in Scripture (Lev. xxiii. 40; Neh. viii. 15). They were used at the Feast of Tabernacles for covering the booths. They were also used as emblems of victory or triumph. Thus palm-leaves, from trees growing on the Mount of Olives, were employed by the multitude when they went forth to meet Jesus coming to Jerusalem (John xii. 13). The custom in England of carrying branches of willow on Palm Sunday seems to have reference to this event. In the heavenly Jerusalem, the great multitude who stood before the throne and before the Lamb are represented by the apostle John as " clothed with white robes, and palms in their hands " (Rev. vii. 9). The flowers are produced on a branching spadix covered by a sheath. The tree having staminate flowers on one plant and pistillate on another, requires to be fertilized by the application of the pollen of the one to the pistil of the other; and unless both kinds are cultivated, the fruit may not be perfected, and the tree may ultimately fail. This may account in part for the disappearance of the palm. The fruit hangs in clusters. This is supposed to be alluded to in Song of Solomon vii. 7 : " This thy stature is like a palm tree, and thy breasts to clusters of dates " (not *grapes*, as given by our translators). Dates constitute an important article of food. It is said that nineteen-twentieths of the population of Fezzan, in Africa, live on dates during nine months of the year; and that many of the animals also feed on them. It is also stated that in Fezzan every door and every post is made of date-palm wood, and that the poorer classes live in huts (booths) entirely made of date-palm leaves. Dates are imported into Britain from Barbary and Egypt, and are usually of the variety called Tafilat.

Figures of palm-trees were introduced by Solomon into the carvings of the temple (1 Kings vi. 29, 32, 35, vii. 36; 2 Chron. iii. 5); and they are also referred to by Ezekiel in his description of the second temple (Ezek. xl. 16, 22, 26, 31, 37; xli. 18–20, 25). The palm-tree, from its erect and noble growth, and its heavenward direction, is used in Psalm xcii. 12 as an illustration of the righteous.

# POMEGRANATE-TREE.

*(Punica Granatum, Linn.)*

---

"Thy plants are an orchard of pomegranates, with pleasant fruits."
Song iv. 13.

THE pomegranate-tree and its fruit are noticed in Scripture under the Hebrew name of *rimmon*. The plant is the *rhoa* of Dioscorides and the *sidé* of Homer. It is a native of Asia, and, according to Royle, may be traced from Syria through Persia and the mountains of Northern India. It was common in Palestine. Thus Moses, speaking of the Promised Land, calls it " a land of wheat and barley, and vines, and fig trees, and pomegranates" (Deut. viii. 8); and the spies who searched the land "brought of the pomegranates and of the figs" (Num. xiii. 23). Several towns and villages in Palestine bore the name of Rimmon or Pomegranate (Josh. xv. 32; 1 Chron. iv. 32, vi. 77; Zech. xiv. 10). Saul tarried under a pomegranate-tree (1 Sam. xiv. 2); and the prophets Joel and Haggai refer to the pomegranate (Joel i. 12; Haggai ii. 19). The tree must have grown in Egypt during the time the Israelites sojourned there; for when in the wilderness of Zin, they lamented the loss of the pomegranates (Num. xx. 5).

The tree is the *Punica Granatum* of botanists, the generic name indicating a Carthaginian origin. The English name pomegranate is derived from the words *pomum granatum,* or grained apple of the Romans; so called from the arrangement of the red seeds in different

POMEGRANATES.

and in separate divisions in the interior of the fruit. The tree belongs to the natural order Lythraceæ, the Loose-strife family. It has a dark green foliage resembling that of the olive and the myrtle; its flowers are of a beautiful crimson colour; and its fruit is red-coloured, as large as an orange, and contains a juicy pulp, which is particularly refreshing in warm countries. The calyx forms part of the fruit. Delicious seedless pomegranates are grown near Cabul.

The beauty of the flower and fruit, and the use of the latter as an article of food, caused the plant to be cultivated in gardens (Song iv. 13; vi. 11; vii. 12). The delicate colour of the pulp of the fruit is referred to in the following passage: "Thy temples [or rather thy cheeks] are like a piece [section] of a pomegranate within thy locks" (Song iv. 3; vi. 7). The pulp of the fruit is eaten alone or with sugar, and the juice is expressed to furnish a refreshing drink, or to form wine. The wine of the pomegranate is mentioned in Song viii. 2.

The pomegranate was selected as a pattern of various ornamental carvings and embroiderings in ancient times. The fruit and the flower furnished beautiful models for the purpose. The chapiters or capitals of the pillars in the temple were covered on the top with carved pomegranates (1 Kings vii. 18, 20, 42; 2 Kings xxv. 17; 2 Chron. iii. 16, and iv. 13; Jer. lii. 22). Embroidered pomegranates, with golden bells between them, were put on the bottom of the high priest's blue robe and ephod (Ex. xxviii. 33, 34, xxxix. 24–26).

Various parts of the pomegranate-tree have been used medicinally, especially for the cure of tape-worm. The bark of the root, the flowers, and the rind of the fruit, have been used for this purpose. The rind was employed for tanning and preparing the finer kinds of leather in early times. It is the principal material used at the present day in the manufacture of morocco leather.

# SHITTAH-TREE.

(*Acacia Seyal, Delile.*)

---

"I will plant in the wilderness the cedar, the shittah tree, and the myrtle, and the oil tree."—ISA. xli. 19.

THE shittah-tree of the Bible is the plant which yielded shittim-wood. This wood is mentioned among the offerings of the children of Israel (Ex. xxv. 5; xxxv. 7, 24). It was used in making the various parts of the tabernacle in the wilderness,—such as the ark and its staves (Ex. xxv. 10, 13, xxxvii. 1, 4; Deut. x. 3); the table for the shew-bread and its staves (Exod. xxv. 23, 28; xxxvii. 10, 15); the boards for the tabernacle and their bars (Ex. xxvi. 15, 26; xxxvi. 20, 31); the pillars for the veil and for the hanging of the door (Ex. xxvi. 32, 37; xxxvi. 36); the altar of burnt-offering and the altar for incense and their staves (Ex. xxxvii. 25, 28; xxxviii. 1, 6).

Considerable differences of opinion have existed relative to the tree which is referred to in these passages of Scripture. It grew apparently in abundance in the

desert, so as to be easily procured by the Israelites.
Dr. Shaw, in speaking of Arabia Petræa, says: "The
acacia-tree, being by much the largest and most common
tree in these deserts, we have some reason to conjecture
that the shittim-wood was the wood of the acacia, especially as its flowers are of an excellent smell; for the
shittah is, in Isaiah xli. 19, joined with the myrtle and
fragrant shrubs." Kitto thinks that the tree is the
*Acacia Seyal* of botanists. This tree belongs to the
natural order Leguminosæ, and sub-order Mimosæ.

The plants of this sub-order of Leguminosæ produce a
legume or pod; and their flowers are regular, their petals
being valvate in æstivation.

The acacia-tree is thorny, and bears pinnate leaves.
Its flowers grow in round yellow clusters, and the long
thread-like projecting stamens give a peculiar character
to the inflorescence. The poet speaks of the acacia as
waving "its yellow hair." Its wood is hard and durable,
and is susceptible of a fine polish. The plant grows in
dry situations, is a native of Egypt, and is scattered
over the whole Sinaitic peninsula. It grows also near
the Dead Sea. It is one of the trees which yield gum
arabic, which is exported in great quantities from the
Red Sea. The gum exudes from the bark, which is
astringent, and is used for tanning. The tree appears
to have grown near Jerusalem, for Joel, in speaking
of the glory of the latter days, says, "And it shall
come to pass in that day that the mountains shall
drop down new wine, and the hills shall flow with
milk, and all the rivers of Judah shall flow with

waters, and a fountain shall come forth of the house of the Lord, and shall water the valley of Shittim;" probably so called from the shittah or acacia trees growing in it (Joel iii. 18). Shittim is also noticed by Micah (vi. 5); and in the journeyings of the children of Israel a place

SHITTAH-TREE.—(*Acacia Seyal.*)

named Abel-shittim is mentioned in the plains of Moab (Num. xxxiii. 49).

It has been supposed that the bush mentioned in Exodus iii. 2–4, and Deuteronomy xxxiii. 16, was a species of acacia, allied to the shittah-tree. It is called in Hebrew *seneh*.

# SYCAMINE-TREE.

(*Morus Nigra,* Linn.; *Black mulberry.*)

---

"If ye had faith...ye might say to this sycamine tree, Be thou plucked up by the root."—LUKE xvii. 6.

THE Greek word *sycaminos*, translated "sycamine-tree," occurs in one passage in the New Testament—namely, in Luke xvii. 6: "And the Lord said, If ye had faith as a grain of mustard seed, ye might say unto this sycamine tree, Be thou plucked up by the root, and be thou planted in the sea; and it should obey you." The tree must not be confounded with the sycamore. It is obvious from Dioscorides, Galen, and other Greek authors, that by sycamine the mulberry-tree was meant. Celsius states this also very distinctly. Sibthorp, who examined carefully the plants of Greece, and published the "Flora Græca," says that in that country the white mulberry-tree is, at the present day, called *mourea*, and the black mulberry-tree, *sycamenia*. Judging, then, from the use of the term at the present day in Greece, it is believed that the *Morus nigra*, or black mulberry, is the species referred to.

Both the white and the black mulberry are common

in Palestine, and are much cultivated, as the leaves supply food for silkworms. The mulberry belongs to the natural order Artocarpaceæ, the Bread-fruit family, and sub-order Moreæ, Mulberry section. The leaves of the black mulberry are large, the flowers are in clusters, and the fruit is the product of many flowers; but though it is thus like a bramble-berry, it is totally different in structure.

Our Lord, in the passage from Luke, refers evidently to some tree which was well known to all his hearers. The mulberry would be a fit tree for such an illustration.

The black or purple, and the white mulberry, are natives of Persia and the adjacent countries. The former produces the best fruit. The latter is the handsomer tree, but it is pruned and lopped for the purpose of furnishing a larger quantity of leaves for the silkworms, which are bred in large quantities in Syria. Lady Callcott says that "in the neighbourhood of Mount Lebanon the land-tax of the peasants is assessed according to the number of mule-loads of mulberry leaves their little farms produce; so that the cultivation of the tree is directed to favour the growth of the leaf, at the expense of the fruit......In the southern part of the Holy Land a palm-tree is usually planted in the court; while towards the north it is replaced by the purple mulberry, the pleasant juice of whose fruit, mingled with water, in which the sweet-scented violet has been infused, forms one of the most grateful kinds of sherbet." (*Scripture Herbal*, 283, 284.) The word translated "mulberry-tree" in the Bible is the name of the trembling poplar, or aspen.

# SYCOMORE-TREE.

(*Ficus sycomorus, Linn.*)

---

"I was an herdman, and a gatherer of sycomore fruit."—AMOS vii. 14.

THE sycomore or sycamore tree of the Bible is quite distinct from that usually called sycamore at the present day in Britain. The latter is a species of maple, and is the *Acer pseudo-platanus* of botanists, often called in Scotland plane-tree. The specific name indicates that the plant has some resemblance to the true plane (*platanus*). This resemblance is seen in the leaves only, for in all other respects the trees are totally different. The sycomore of Scripture, however, is a kind of fig-tree, producing fruit similar in structure to the common fig, and having leaves like the mulberry. Hence the name sycomore, which is derived from *sycon*, a fig, and *moron*, a mulberry. It is the *Ficus sycomorus* or the *Sycomorus antiquorum* of botanists. In Hebrew the sycomore-trees are called *shikmoth* and *shikmim*. These are two plural words which occur in several places in the Old Testament. In the New Testament the plant is mentioned under the Greek name of *sycomoros*. The tree belongs to the natural order Artocarpaceæ, the Bread-fruit family; which by some is considered a sub-division of the Urticaceæ, the Nettle family. It is separated from the latter family by its milky juice and the nature

SYCOMORE-TREE.—(*Ficus sycomorus.*)

of its fruit, which is formed by numerous flowers on an elongated or hollow receptacle. The juice usually contains caoutchouc, and the fruit is generally edible.

The sycomore-fig was common in the plains of Egypt, and in the valleys of Palestine. Hence it has been sometimes called Pharaoh's fig; and it is said that Solomon made cedar trees "to be as the sycomore trees that are in the vale for abundance" (1 Kings x. 27; 2 Chron. i. 15, ix. 27). It is still cultivated near Cairo for its shade. It was not valued much either for its timber or for its fruit. Isaiah represents Ephraim and the inhabitants of Samaria as saying in the pride and stoutness of their heart, "The sycomores are cut down, but we will change them into cedars" (Isa. ix. 10); or, in other words, in place of houses built with the common sycomore fig-tree, we will build palaces of cedar. The wood of the sycomore is coarse-grained. In Egypt, where there were few native trees of value, the timber was used to form mummy cases. On account of the dry climate of that country, and the means used for the preservation of the timber, the wood of these cases is very durable.

The fruit of the sycomore grows in clusters on the trunk and main branches. It is edible, and is hence mentioned along with the olive and the vine as one of the products of Canaan,—parties being appointed to take care of the trees (1 Chron. xxvii. 28; Ps. lxxviii. 47). It has a sweetish taste, and is still used as food. It is said to furnish a considerable portion of the food of the field labourers in Rhodes, Cyprus, and Egypt. In order that the fruit might ripen well and be palatable, it was necessary to make incisions into it or to scrape off a part

SYCOMORE FIGS.

at the end of it; and this practice is supposed to be alluded to by Amos when he says, "I was an herdman, and a gatherer of [*literally*, one who scraped or cut] the sycomore fruit" (Amos vii. 14). This mode of fig-ripening is noticed by Pliny. The fruit of this tree might be referred to by Jeremiah when he saw in the vision a second basket—"very naughty figs, which could not be eaten, they were so bad" (Jer. xxiv. 2). The tree was lofty and shady, and hence probably was planted along the road-sides. The stem sometimes attains fifty feet in circumference. Into a sycomore-tree Zaccheus climbed to see Jesus, on that memorable occasion when salvation came to him and to his house (Luke xix. 4). It is easy to climb, as it has a short trunk, dividing into forking branches, which fork out in all directions.

# TEIL-TREE, OR TEREBINTH-TREE.

(*Pistacia Terebinthus, Linn.*)

---

"As a teil tree [terebinth-tree], or as an oak, whose substance is in them."
Isa. vi. 13.

THE feminine Hebrew word *elah* or *ailah*, denoting a strong hardy tree, occurs in several passages of the Bible, and has been variously translated. It is rendered in different versions terebinth, teil-tree, elm, oak, and plain. The word also occurs in the masculine form as *el* or *ail*. It is now generally assumed that the plant indicated is the terebinth-tree, the *buthma* of the Syriac or Chaldee, the *butm* or *botom* of the Arabs, and the *Pistacia Terebinthus* or turpentine-tree of botanists. It belongs to the natural order Anacardiaceæ or Terebinthaceæ, the Cashew family, the plants belonging to which abound in a resinous or milky acrid juice.

The tree is the source of the Chian turpentine, which is procured by incisions in the trunk, and is collected chiefly in the island of Scio: a single tree yields about ten ounces. It is common in Palestine. Dr. Robinson states that the tree is found also in Asia Minor (many near Smyrna), Greece, Italy, the south of France, Spain, and in the north of Africa; and that it sometimes attains the height of thirty or thirty-five feet. He noticed a very large specimen between Gaza and Jerusalem. The tree appears to be long-lived, and was consequently frequently employed to designate places where important

Teil-tree, or Terebinth-tree.—(*Pistacia Terebinthus.*)

events occurred. The favourite burying-place of a Bedouin sheik is under a solitary terebinth-tree.

The valley of Elah or the Terebinth valley is mentioned in 1 Samuel xvii. 2, 19; xxi. 9. It was by this valley that Israel encamped, and it was in this valley that David slew Goliath. In Genesis xiv. 6, El-paran is noticed. This is rendered by the Septuagint "the terebinth of Paran;" by some commentators it is called "the oak of Paran," and by others "the plain of Paran," which is given in our Bibles as a marginal reading. In other places the word is also translated "plain." This variety of translation has given rise to much confusion.

It would appear, also, that the name has been con-

BRANCH OF TEIL-TREE, OR TEREBINTH-TREE.

founded with *allon* and the feminine *allah*, which mean "oak." The difference between the words is well seen in some passages where both occur. Thus in Isaiah vi. 13 it is said, "As a teil tree [elah, or terebinth-tree] and an oak [allon]." So also in Hosea iv. 13, "They sacrifice upon the tops of the mountains, and burn incense upon the hills, under oaks [allon], poplars [libneh], and elms

[elah]." The term "oak" is used instead of terebinth in many other passages, such as the following:—The angel appeared to Gideon under a terebinth at Ophrah (Judges vi. 11, 19); idols were worshipped in groves of terebinth (Isa. i. 29; Ezek. vi. 13); idolaters are compared to a terebinth whose leaf fadeth (Isa. i. 30). See also 1 Kings xiii. 14; 1 Chron. x. 12. Abraham's oak at Mamre was also a terebinth-tree. (See *Oak*.) In figuring the restoration of the mourners in Zion, Isaiah says, "That they might be called trees [terebinths] of righteousness, the planting of the Lord, that he might be glorified" (Isa. lxi. 3).

# HUSK-TREE.

### (*Ceratonia Siliqua, Linn.*)

---

"The husks that the swine did eat."—LUKE xv. 16.

THE Greek word *keratia*, or *ceratia*, occurs in Luke xv. 16, and has been translated "husks." The prodigal son, it is said, "would fain have filled his belly with the husks that the swine did eat." In Arabic, the word is rendered *charnub*, or *charub*, which seems to refer to the pods or legumes of the carob-tree, *caroba* of the Italians, *algaroba* of the Spaniards and Moors, *Ceratonia Siliqua* of botanists. The tree belongs to the natural order Leguminosæ, the legume-bearing family, and section Cæsalpineæ, in which the petals have a pea-like arrangement, but the upper one is interior.

The tree is common in the south of Europe as well as in Syria and Egypt. Its pods or husks received the name of *keratia* from their fancied resemblance to a slightly curved horn, or *keras*. These husks were formerly used in large quantity to feed cattle and swine, and they are often mentioned in this point of view by old authors. Horace, in his Epistles, alludes to living upon husks as upon vile food—

"Vivit siliquis et pane secundo."—EPIST. II. i. 123.

Persius and Juvenal also allude to them. Pliny describes

HUSK-TREE.—(*Ceratonia Siliqua.*)

them as the food of pigs (lib. xv., cap. 23, 24). At the present day, they are employed in Spain and other countries to feed horses, asses, and mules; and they were frequently given to horses by the British soldiers during the Peninsular War. The pods are imported into Britain in small quantity, as food for horses and cattle. The locust beans, as they are called by farmers, are mixed with oil-cake and a little meal. They do not require to be crushed; for, being very palatable, the animals masticate them well before swallowing them. Camels are also fed on them. Hence they are called by the Turks *deweh etmeghi*, or "the bread of the camel." A tree will sometimes produce eight hundred to nine hundred pounds of pods.

The pod is six to eight inches in length, and about an inch in breadth. It is flattened on the sides, and is about a quarter of an inch in thickness. The seeds are of a reddish brown colour, and are immersed in a sweetish pulp. In times of scarcity, the pod has been used by man as food. Some have called the tree locust-tree, and St. John's bread-tree, from a mistaken notion that its pods were the locusts referred to in Matt. iii. 4, and Mark i. 6, as forming part of the food of the Baptist. The German name for the fruit, for the same reason, is *Johannisbrod*.

POD, LEAVES, AND FLOWER OF HUSK-TREE.

Rawolf, in his account of a journey from Bethlehem to Jerusalem, says: "Along the roads were a good many of the trees which are called by the inhabitants *chernubi* (the Arabic *charnub*), and the fruit of which we call St. John's bread; it was brought to us in great quantities."

In the case of the prodigal son, the feeding on husks pointed out the low and miserable condition to which he was reduced when he wandered from his father's house. He would fain have been content with the most miserable fare, and was in a very degraded situation, although, in his madness and folly, he knew it not. His condition represents that of the sinner who has wandered from God, and who is content with the unsatisfying husks of this world's enjoyments.

# PLANE-TREE.

(*Platanus orientalis, Ait.*)

---

"The chesnut trees [plane-trees] were not like his branches."—EZEK. xxxi. 8.

IN two passages of the Old Testament we meet with the Hebrew word *armon*, and in both of them it has been translated "chesnut." Thus, in Genesis xxx. 37, it is said, "And Jacob took him rods of green poplar, and of the hazel and chesnut tree;" and again, in Ezekiel xxxi. 8, "The cedars in the garden of God could not hide him; the fir trees were not like his boughs, and the chesnut trees were not like his branches." The best commentators consider the tree to be the Eastern plane-tree, the *Platanus orientalis* of botanists. It is so rendered in the Septuagint. It is a large tree, with spreading branches. Ovid speaks of "platano conspectior alta;" and Martial alludes to the tree thus: "Ramis sidera celsa petit." De la Roque, in his "Travels in Syria and Mount Lebanon," says: "We dined in the midst of this little forest. It is composed of twenty cedars, of such enormous size that they far exceed the more beautiful

plane-trees, sycomores, and other large trees, which we had been in the habit of seeing during our journey." Royle says: "It may be remarked that this tree is in Genesis associated with such trees as the willow and poplar, which, like it, grow on low grounds, where the soil is rich and humid. Russel names the plane, willow,

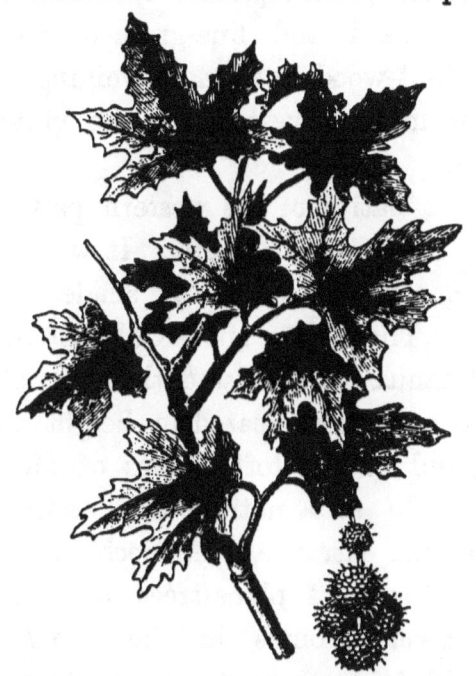

PLANE-TREE.—(*Chesnut-tree of Scripture.*)

and poplar as trees which grow in the same situations near Aleppo. This congruity would be lost if the chestnut were understood, as that tree prefers dry and hilly situations."

The plane-tree belongs to the natural order Platanaceæ, or the Plane-tree family, which are catkin-bearing plants, with the flowers in clusters of rounded balls, pendulous on a common stalk. The leaves of the Oriental plane are palmate, resembling those of our common sycamore, which is a species of maple. The resemblance in the form of the leaves has caused the latter to be denominated in Scotland the plane-tree, and to be named botanically false plane (*Acer pseudo-platanus*). The wood of the true plane is hard and fine-grained, and when old resembles walnut-wood in its dark veining. The timber was used for making vessels for the vintage, and for other purposes.

The tree is a native of the western parts of Asia, and it extends as far as to Cashmere. It was held sacred in the East, and was valued for its shade by the Greeks and Romans. Themistius speaks of disputations under the lofty platanus. Belon says that the plane-trees of Mount Athos may be compared in height to the cedars of Lebanon, and to the lofty pines of Mount Olympus and Aman. He also notices the occurrence of fine plane-trees at the entrance to Antioch; and De la Roque refers to the forest of plane-trees and cypresses which border the river Orontes, in the plains of Antioch. Xerxes is said to have paid homage to a large plane-tree in Lydia.

# NUTS;

INCLUDING THE WALNUT TREE (*Juglans regia, Linn.*) AND
THE PISTACIA NUT (*Pistacia vera, Linn*).

---

"The garden of nuts" [walnuts].—SONG OF SOLOMON vi. 11.
"Carry down...myrrh, nuts [pistacia nuts], and almonds."—GEN. xliii. 11.

THE Hebrew word *egoz* has been rendered, in our version of the Bible, "nuts." It occurs in the Song of Solomon (vi. 11): "I went down into the garden of nuts to see the fruits of the valley." It is the Arabic *gjaus* or *ghaus*, and the Syriac *gusa*, which were names given to the walnut. Hence the plant is believed to be the walnut-tree, the *Juglans regia* of botanists. The fruit is the *caruon basilicon* or royal nut of the Greeks, the *nux* of the Romans, and the *noix* of the French. The Latin term *Juglans* is a corruption of Jovis-glans, or Jupiter's nut. It appears to have been one of the many kinds of fruits which Solomon introduced into his gardens and orchards (Eccles. ii. 5).

The walnut-tree belongs to the natural order Juglandaceæ, the Walnut family, in which the flowers are in catkins, and the fruit is a drupe, usually with a two-valved endocarp or shell, and a peculiarly lobed and divided seed. The latter character is well seen in the common walnut. The tree is wide-spreading, and affords a grateful shade. It flowers in April, and has ripe fruit in September and October. Its leaves are fragrant when bruised. The outer covering of the fruit is astringent, and dyes the fingers black during the process of peeling.

WALNUT-TREE.—(*Juglans regia.*)

The thin outer covering of the seed immediately under the shell is bitter, and in its fresh state requires to be removed before the kernel is eaten. The seed yields a large quantity of drying oil. The timber is valued for carpenter-work.

The tree extends from Greece and Asia Minor through Persia to the Himalaya. In Cashmere walnuts are cultivated for their oil. Josephus says that the walnut-trees were very productive around the Lake of Gennesaret. Schulz also mentions large walnut-trees between Ptolemais and Nazareth. Travellers record the occurrence of the tree in Syria; Thevenot found it near Mount Sinai, and Belon alludes to it as abundant near Lebanon.

Another Hebrew word, *botnim,* has been also rendered "nuts" in our version of the Bible. It occurs in Genesis xliii. 11, where Israel says to his sons, "Take of the best fruits in the land in your vessels, and carry down the man a present, a little balm, and a little honey, spices, and myrrh, nuts, and almonds." Various plants have been considered as yielding the nuts referred to in this passage. Considering that the fruit was the common produce of Syria, and that the allied Arabic word, *batam* or *botin,* is applied to a species of terebinth, it is now supposed that Bochart was correct in saying that the nuts were the pistacia or pistachio nuts of commerce, the produce of the *Pistacia vera* of botanists. *Betonim,* a name applied to a town of the Gadites (Josh. xiii. 26), is probably a modification of the same word.

The pistacia-nut tree belongs to the natural order Anacardiaceæ or Terebinthaceæ, the Cashew family. The green-coloured kernels yield oil. Royle says: "Pistachio nuts are much eaten by the natives of the countries where they are grown, and they form an article of commerce from Afghanistan to India. They are also exported from Syria to Europe. They might, therefore, well have formed a part of the present intended for Joseph." Aleppo is still famous for pistachio nuts.

# VINE.

### (*Vitis vinifera, Linn.*)

---

"I am the true vine, and my Father is the husbandman."—JOHN xv. 1.

THE vine is expressed in Hebrew by the word *gephen*, and in Greek by the word *ampelos*; while the grape or the fruit of the vine is the *anul* or *yayin* of Hebrew, and the *staphylé* of Greek writers. The plant is called by botanists *Vitis vinifera*. It belongs to the natural order Vitaceæ or Ampelideæ, the Vine family.

The vine, its fruit, and the wine made from it, are often referred to in the Bible. The plant is said to be a native of the hilly region on the southern shores of the Caspian, of the Persian province of Ghilan, and of Armenia. It has been distributed extensively over the world, and its cultivation is noticed in the earliest times. Noah planted a vineyard after the Deluge, probably in Armenia, and made wine from the grapes (Gen. ix. 20, 21). Wine is mentioned in the interview between Abraham and Melchizedek (Gen. xiv. 18). In the blessing of Judah, Jacob says, "He washed his garments in wine, and his clothes in the blood of grapes" (Gen. xlix. 11), as indicative of prosperity. The vine was known to the Egyptians (Gen. xl. 9–11), and is represented on their monuments. The Israelites, in their journey through the wilderness, longed for the vines of Egypt (Num. xx. 5); and the psalmist, in alluding to God's judgments on Pharaoh, introduces the vines as being destroyed (Ps. lxxviii. 47; cv. 33).

Vineyards abounded in Canaan when the Israelites took possession of it, and the vines were very productive. The Promised Land seems specially the country of the vine, and the climate is still fitted for its cultivation. The evidences of terraced vineyards, of wine-presses and vats, still remain. The men who were sent by Moses to search the land, cut in *Nachal-Eshcol*—that is, the Valley of Eshcol, or the Grape Valley, near Hebron—a cluster of grapes which was so large that it was carried by two upon a staff (Num. xiii. 23). In Syria, at the present day, clusters weighing ten or twelve pounds have been gathered. Frequent allusions occur in the Bible to vineyards, to vine-dressers, to the rejoicing at the vintage, the gathering and the gleaning of grapes, the treading of the grapes, the wine-presses and the wine-fats—all indicating the important place which the vine occupied among the vegetable productions of Palestine. In that country the vintage extended from the beginning of September to the end of October. In speaking of the future prosperity of Israel, Amos says that the days will come when "the treaders of grapes shall overtake him that soweth seed" (Amos ix. 13), thus indicating a long-continued vintage; while, in speaking of desolation, Isaiah says, "In the vineyards there shall be no singing, neither shall there be shouting: the treaders shall tread out no wine in their presses; I have made their vintage shouting to cease" (Isa. xvi. 10).

The treading of the grapes is often mentioned as a joyful occasion: "He shall give a shout, as they that tread the grapes" (Jer. xxv. 30). The treaders had their feet and legs bare, but when first leaping on the grapes the juice often dyed their clothes. Hence the allusion in Isaiah lxiii. 2, 3, where the Lord speaks of treading the wine-press alone, and of staining all his raiment.

In gathering the grapes, the Israelites were told to leave gleanings for the poor and the stranger (Lev. xix. 10).

Some choice vines are mentioned under the name of *sorek* (Gen. xlix. 11; Isa. v. 2; Jer. ii. 21). The vineyards of Eshcol, Heshbon, Elealeh, Sibmah, Jazer, Engedi, and Helbon, were celebrated (Song of Sol. i. 14; Isa. xvi. 8–10; Jer. xlviii. 32, 33; Ezek. xxvii. 18).

The wine of Helbon is still famous; and Damascus must always have been the natural channel for its export. Ezekiel says: "Damascus was thy merchant in the multitude of the wares of thy making, for the multitude of all riches; in the wine of Helbon, and white wool." The "wine of Lebanon" is also mentioned in Hosea xiv. 7, and it is still in repute.

Vineyards were specially protected. In the Mosaic law there was an injunction against trespassing on vineyards. Isaiah thus speaks of a vineyard: "My well-beloved hath a vineyard in a very fruitful hill: and he fenced it, and gathered out the stones thereof, and planted it with the choicest vine, and built a tower in the midst of it [for watchers], and also made a winepress therein" (Isa. v. 1, 2). In Psalm lxxx. 12, allusion is made to the hedges of the vineyard; and in the New Testament it is said, "A householder planted a vineyard, and hedged it round about, and digged a winepress in it, and built a tower" (Matt. xxi. 33; Mark xii. 1). Jackals (called foxes in Scripture) were apt to injure the vines. Hence, in the Song of Solomon, we have, "Take us the foxes, the little foxes, that spoil the vines: for our vines have tender grapes" (ii. 15). Vines were trained in various ways in Palestine. Sometimes they were trained over walls or trellises, so as to form a complete bower almost like a tree. Perhaps the patriarch Jacob alluded to this in the blessing of Joseph, when he spoke of "a fruitful bough, whose branches run over the wall" (Gen. xlix. 22); and the psalmist when he speaks of "a fruitful vine by the sides of thine house" (Ps. cxxvii. 3).

The vine has followed the footsteps of man, and has been transplanted by him into all parts of the world. The juice of the young fruit, called *verjuice*, is very sour; that of the riper fruit is called *must*, and is used as a refreshing drink in some countries. It is probably referred to in those passages of Scripture in which the expression "the blood of the grape" occurs. There was a thin sour wine used by the poorer classes, which is often translated "vinegar" (Ruth ii. 14). It might be this which was offered to the Saviour on the cross (Matt. xxvii. 48). The dried fruit, known as "raisins," is also noticed in the

Bible (1 Sam. xxv. 18, xxx. 12; 2 Sam. xvi. 1; 1 Chron. xii. 40). Many illustrations are taken from the vine. Israel is represented as a vine brought from Egypt, and planted by the Lord (Ps. lxxx. 8–11; Isa. v. 7; Jer. ii. 21). Dwelling under the vine and fig-tree is an emblem of peace and tranquillity (Micah iv. 4; Zech. iii. 10). A fruitful vine is associated with domestic happiness (Ps. cxxviii. 3). The production of wild grapes and of grapes of gall, an empty vine, and a strange vine, are figurative expressions used to illustrate the departure of Israel from God (Deut. xxxii. 32, 33; Isa. v. 2, 4; Jer. ii. 21; Hos. x. 1). The phrase "wild grapes" is by some translated "putrid grapes." They are considered by Berkeley as grapes affected with rot or mildew. Our Saviour calls himself the true vine, into which his disciples are grafted, so as to bring forth much fruit (John xv.).

Robinson mentions an ancient wine-press at Hablch. "Advantage had been taken of a ledge of rock; on the upper side, towards the south, a shallow vat had been dug out eight feet square and fifteen inches deep, its bottom inclining slightly towards the north. The thickness of the rock left on the north was one foot, and two feet lower down on that side another smaller vat was excavated four feet square by three feet deep. The grapes were trodden in the shallow upper vat, and the juice drawn off by a hole at the bottom into the lower vat." (*Biblical Researches in Palestine,* 1856, p. 137.) In Isaiah xvi. 7, it is written, "For the foundations of Kir-hareseth shall ye mourn;" but it should be, "For the raisin-cakes of Kir-hareseth shall ye mourn." (See Hosea iii. 1.)

# WILLOW-TREE.

(*Salix babylonica, Linn.*)

---

"They shall spring up...as willows by the water courses."—ISA. xliv. 4.

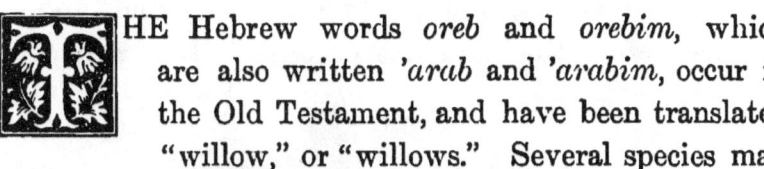HE Hebrew words *oreb* and *orebim*, which are also written *'arab* and *'arabim*, occur in the Old Testament, and have been translated "willow," or "willows." Several species may have been included under the name *orebim*. We have figured *Salix babylonica*, the weeping-willow, as being probably one of them, and as being that more especially referred to in Psalm cxxxvii. 1, 2, when Israel in captivity says, "By the rivers of Babylon, there we sat down; yea, we wept, when we remembered Zion. We hanged our harps upon the willows in the midst thereof."

Willows belong to the natural order Salicineæ, the Willow family, consisting of useful timber trees having a tonic and astringent bark, flowers in catkins, and seeds covered with silky hairs.

Willows are found in moist situations, beside running brooks as well as by still waters. In the Bible, the locality of their growth is usually associated with them. Thus, on the first day of the Feast of Tabernacles, the Israelites are enjoined to take "boughs of goodly trees, branches [leaves] of palm trees, and the boughs of thick trees, and willows of the brook" [wady or ravine], and to "rejoice before the Lord...seven days" (Lev. xxiii. 40).

WILLOW-TREE.—(*Salix babylonica.*)

These were employed in the construction of booths (Lev. xxiii. 42). Job, in describing behemoth (probably the hippopotamus), says: "The shady trees cover him with their shadow; the willows of the brook compass him about" (Job xl. 22). In the seventeenth verse of the same chapter, the word "cedar" ought to be *willow*— "He moveth his tail like a willow." In proclaiming the burden of Moab, the prophet says: "Therefore the abundance they have gotten, and that which they have laid up, shall they carry away to the brook [valley] of the willows" (Isa. xv. 7). Again, in comforting the Church with his gracious promises, God speaks thus by the mouth of his prophet: "And they [their offspring] shall spring up as among the grass, as willows by the water courses" (Isa. xliv. 4); indicating a constant supply of refreshing water, when the Lord "will pour water upon him that is thirsty, and floods upon the dry ground" (Isa. xliv. 3).

Willows were thus associated both with the joyous and the sorrowful days of the children of Israel. When captives in Babylon, their grief was poured forth under the willows; and in contemplating God's purposes of mercy towards them, they are directed to the willows as emblems of their growth, and as recalling the willows of the brook with which they rejoiced in their feast-days of old.

Another Hebrew word, *tzaphtzapha*, or *zaphzapha*, has been translated "willow-tree." It occurs in Ezekiel xvii. 5: "He took also of the seed of the land, and planted it in a fruitful field; he placed it by great waters, and set it as a willow tree." This appears to be a species of willow called by the Arabs *safsaf*, which is the generic name. This may be *Salix ægyptiaca* of botanists. This tree was noticed by Hasselquist in his journey from Acre to Sidon.

There are other species of willow in the Holy Land, such as *Salix octandra*, *Salix viminalis*, the common osier. The word *safsaf* is found in the names of places. Thus there is Wady Safsaf, "the valley of the willow;" and Ain Safsaf, "the fountain of the willow."

Tristram, from his observations in Palestine, is disposed to think that the "willow by the water-courses" is

OLEANDER.

more especially applicable to *Nerium Oleander*, the rose bay oleander. This is a native plant of Palestine, and is a very conspicuous one, and could hardly be omitted in the Bible. "It fringes the whole Upper Jordan, dipping its wavy crown of red into the spray in the rapids under Hermon, and is nurtured by the oozy marshes in the Lower Jordan nearly as far as to Jericho. The immediate basin of the Dead Sea is too hot for it; everywhere else it demands but moisture, and springs up by the water-courses." This plant, the *Nerium Oleander* of Linnæus, belongs to the natural order Apocynaceæ, the Dog-bane family, which contains many poisonous plants. Some think that it corresponds to the Rhodon, or rose of the Apocrypha.

# CAMPHIRE.

(*Lawsonia inermis, Linn.*)

---

"As a cluster of camphire in the vineyards of En-gedi."—SONG OF SOL. i. 14.

THE Hebrew word *kopher* or *copher* occurs in the Song of Solomon, and has been translated "camphire." Thus the king says, "My beloved is unto me as a cluster of camphire in the vineyards of En-gedi" (i. 14); and again, "Thy plants are an orchard of pomegranates, with pleasant fruits; camphire, with spikenard" (iv. 13). The Hebrew word resembles the Greek *kupros* or *cypros*, which is applied by Dioscorides and Pliny to a plant known to botanists by the name of *Lawsonia inermis*. It belongs to the natural order Lythraceæ, the Loose-strife family. It is an odoriferous shrub, the *henna* or *alkanna* of Cyprus and Egypt. It is found in Arabia, and on the shores of the Dead Sea at En-gedi. Its fragrant flowers grow in clusters, and it is used in the East for dyeing the nails, the palms of the hands, and the soles of the feet of an iron-rust colour. The object of this custom is to check perspiration. Henna powder is procured from the stem and leaves of the plant. These are put into hot water, stirred and boiled well, and then left on the fire for two hours, until the mass becomes a paste, which when applied to the hair and the skin, tinges them with an orange colour. The plant is used for dyeing morocco leather. The custom of dyeing the nails was an ancient one in Egypt. It is said that the nails of mummies (especially females) have sometimes traces of it. Some think that there is an allusion made to the practice in Deuteronomy xxi. 12, where, in place of "pare her nails," the phrase might be rendered, "adorn her nails."

*In addition to the Plants which we have figured and described, there are other Trees and Shrubs which are briefly or very obscurely alluded to in the Bible. The following are recorded here in order to complete the list.*

## ALMUG OR ALGUM TREE.

(*Santalum album*, Linn., according to some, while according to others it is *Pterocarpus santalinus*.)

" The navy of Hiram...brought in from Ophir great plenty of almug trees."
1 KINGS x. 11.

ALGUM or almug trees are mentioned in Scripture; they do not appear, however, to have been trees of the Holy Land.

The Hebrew words *almuggim* and *algummim* are translated "almug" or "algum trees," in our version of the Bible. The plant referred to is supposed to be the sandal-wood of India called *Santalum album* by botanists, and belonging to the natural order Santalaceæ, or the Sandal-wood family. Others, however, think that the plant is more probably *Pterocarpus santalinus*, or the red sandal-wood of India. This belongs to the natural order Leguminosæ and the sub-order Papilionaceæ. The wood was brought from Ophir (probably some part of India) by Hiram, and was used in the formation of pillars for the temple, and for the king's house, as well as for harps and psalteries (1 Kings x. 11, 12 ; 2 Chron. iii. 8, ix. 10, 11). The wood of the white sandal-wood is fragrant, and is used for incense in China. Large quantities of this sandal-wood are cut in Malabar for export to China and different parts of India. The outer wood of the stem is white and has no odour, while the central part, especially near the root, is fragrant. The red sandal-wood is a heavy, fine-grained, red-coloured wood, and is used in the East for making musical instruments.

## ALOES-TREE, OR LIGN-ALOES TREE.

### (*Aquilaria Agallochum, Roxb.*)

"All thy garments smell of myrrh, aloes, and cassia."—PSALM xlv. 8.

THE Hebrew words, *ahálim* and *aháloth*, and the Greek *aloe*, are rendered "aloes," in our version of Scripture. The substance is supposed by some to have been the fragrant wood of *Aquilaria Agallochum*, by others to be the gum. This is a plant belonging to the natural order Aquilariaceæ, found in Northern India and Cochin China, and attaining a height of one hundred and twenty feet, but which does not occur in Syria or Chaldæa. There is, therefore, some doubt as to the tree mentioned in Scripture. Trees of lign-aloes are referred to in Numbers xxiv. 6. The use of aloes as a perfume is noticed in Psalm xlv. 8 ; Proverbs vii. 17 ; in the Song of Solomon iv. 14 ; and for perfuming the coverings of the dead it is referred to in John xix. 39, 40, where it is said that Nicodemus, after the manner of the Jews, "brought a mixture of myrrh and aloes, about an hundred pound weight," in order to impart fragrance to the linen clothes in which our Saviour's body was wound. We must not confound this aloes with the bitter aloes so well known as a medicine, which is the produce of a totally different plant, and which does not possess the fine fragrance of the substance now under consideration.

## ASH-TREE.

"He planteth an ash [oren], and the rain doth nourish it."—ISA. xliv. 14.

THE Hebrew word *oren*, which occurs in Isaiah xliv. 14, is translated "ash" in our version. It is supposed by some to mean a kind of pine-tree, while others look upon it as a thorny shrub allied to Rhamnus or Capparis. We still want information on the subject. The Syrian pine is *Pinus halepensis*. It belongs to the natural order Coniferæ. Tristram says that there is a tree in the valleys of Arabia Petræa, called in Arabic *aran*, whose foliage resembles that of our mountain-ash.

## EBONY-TREE.

(*Diospyros ebenus, Retz.*)

"They brought thee for a present horns of ivory and ebony."
EZEK. xxvii. 15.

THE Hebrew word, *hobnim*, occurs in Ezekiel xxvii. 15, and has been translated "ebony." It was brought to Tyre by the merchants of Dedan, who came from the Persian Gulf. This wood appears to be the product of various

EBONY-TREE.—(*Diospyros ebenus.*)

trees, more particularly of species of Diospyros—such as *Diospyros ebenus.* They belong to the natural order Ebenaceæ, and are valued for their hard and durable timber. The outside wood of the ebony-tree is white and soft, while the central part is black and hard. The ebony-tree is a native of Ceylon and Southern India.

---

## JUNIPER-BUSH.

### A KIND OF BROOM.

*(Genista monosperma, Lam.)*

" He lay and slept under a juniper tree [rothem]."—1 KINGS xix. 5.

THE Hebrew word *rothem*, or *rotem*, has been rendered "juniper" in our version. It seems to be the same as the Arabic word *retem*, and the *retama* of the Moors. These terms are applied to a kind of broom. It is believed that *rotem* is the *Genista monosperma* of botanists, belonging to the natural order Leguminosæ (Pea family), and includes Papilionaceæ. It is a shrubby plant with white blossoms. It is found in Spain, Portugal, Barbary, Egypt, Syria, and Palestine. Elijah rested under the shade of the rotem or broom (1 Kings xix. 4, 5). Lord Lindsay states that, during his travels in the valleys of Mount Sinai, "the *rattam*, a species of broom, bearing a white flower delicately streaked with purple, afforded him shelter from the sun while in advance of the caravan." Dean Stanley also mentions that during a storm of rain in the desert he sheltered himself under a retem bush. It is a large and conspicuous plant in deserts. It is used for making charcoal by the Arabs. The use of the plant as fuel is referred to in Psalm cxx. 4, " Sharp arrows of the mighty, with coals of juniper ;"

and it would appear that its roots were eaten in certain circumstances, for Job says, "Who cut up mallows by the bushes, and juniper roots for their meat" (Job xxx. 4). The Israelites encamped in the wilderness of Paran at Rithmah, or "place of broom" (Num. xxxiii. 18).

## POPLAR.

### (*Populus alba, Linn.*)

"They burn incense under oaks, and poplars, and elms."—Hos. iv. 13.

THE Hebrew word *libneh,* meaning "white," has been translated "poplar" in Genesis xxx. 37 and Hosea iv. 13. Some think that the name may refer to the white poplar (*Populus alba* of Linn.), which is remarkable for the whiteness of the under side of its leaves. This tree belongs to the natural order *Salicineœ,* the Willow family. The tree grows in the mountainous districts of Palestine, and in various parts of Galilee, Lebanon, and Mount Hermon. On the banks of the Jordan another species of poplar grows (*Populus euphratena*), the Euphrates poplar. Layard says that the only trees within the land of Assyria sufficiently large to furnish beams to span a room thirty or forty feet wide are the palm and the poplar. Their trunks still form the roofs of houses in Mesopotamia. The boats now employed in the lower parts of the Tigris and Euphrates are constructed of planks taken from a species of poplar, probably *Populus euphratena.*

Some authors think that the word *libneh* refers to

the storax plant (*Styrax officinale* of Linnæus), which has white flowers, and a white lower surface of the leaves. The plant yields a fragrant resin called storax, containing benzoic acid, and is used as a pectoral remedy. Storax is probably the stacte of the Bible. (See *Stacte*.)

## MYRRH-TREE.

(*Balsamodendron Myrrha, Ehrenb.*)

"Perfumed with myrrh and frankincense."—Song of Sol. iii. 6.

Myrrh.

The Hebrew word *mohr* and the Greek *smurna* are translated "myrrh" in the Bible. This substance is a fragrant sort of gum which exudes from various trees in Arabia and Abyssinia, one of the chief being *Balsamodendron Myrrha*, or the myrrh balsam-tree, belonging to the natural order Amyridaceæ, the Myrrh family. Myrrh was celebrated as a perfume, and as a stimulant medicine. It was burned in temples, and was employed in embalming (John xix. 39). It entered into the composition of the holy anointing oil (Ex. xxx. 23). It was given as a present, from its value and rarity (Matt. ii. 11), and its fragrance is often made mention of (Ps. xlv. 8; Song of Sol. iii. 6, iv. 6, 14, v. 1, 5, 13).

The balm of Gilead is supposed to be the produce of *Balsamodendron gileadense.* The native country of the plant is the east coast of Africa, but it was cultivated extensively in the plains of Jericho. From the bark of the tree there is obtained by incision a fragrant exudation of a yellow colour, which was used as a stomachic, and was applied to wounds.

Many species of Balsamodendron are called balsam-trees. They are mentioned under the Hebrew names of *basam* and *Baal-shemen.* The word *tzeri,* also translated "balm," occurs in Gen. xxxvii. 25, xliii. 11; Jer. viii. 22, xlvi. 11, li. 8; and Ezek. xxvii. 17. "Is there no balm in Gilead? is there no physician there?" (Jer. viii. 22.) The word *basam* is often translated "spices" (Song of Sol. v. 1, 13, vi. 2; Ex. xxxv. 28; 1 Kings x. 10). (See also *Ladanum.*)

## ESHEL.

*(Translated Grove, and Tree.)*

"Saul abode in Gibeah under a tree [eshel or tamarisk] in Ramah."
1 SAM. xxii. 6.

ESHEL is a Hebrew word, which occurs in Genesis xxi. 33, where it is translated "grove;" and in 1 Samuel xxii. 6 and xxxi. 13, where it is translated "tree." It is said that "Abraham planted a grove [eshel] in Beer-sheba, and called there on the name of the Lord;" that "Saul abode in Gibeah under a tree [eshel] in Ramah;" and finally, that Saul and his sons were buried " under a tree [eshel] at Jabesh." Royle considers the word as being

TAMARISK-TREE.

equivalent to the Arabic *asul* or *athul*, which refers to a large species of tamarisk; and he regards *eshel* as

*Tamarix orientalis*, the Eastern tamarisk-tree. Bové mentions the tamarisk as occurring in Syria. It is a tree which thrives in arid, sandy situations. It belongs to the natural order *Tamariscineæ*, the Tamarisk family. *Tamarix pallasia* grows in the arid sands of the Dead Sea; and *Tamarix gallica* is also found near the shores of Palestine.

## THYINE-WOOD.

(*Xylon thyinum.*)

"Merchandise of silk, and scarlet, and all thyine wood."—Rev. xviii. 12.

THYINE-WOOD is mentioned in Revelation xviii. 12, as one of the articles of merchandise in the Apocalyptic Babylon. This appears to be the *citron-wood* of the Romans, the *alerce* of the Moors, the *Thuja articulata* of Vahl, and the *Callitris quadrivalvis* of Ventenat. This tree, called also the *arar-tree*, belongs to the natural order Coniferæ, or Cone-bearers, and the sub-order Cupressineæ, or the Cypress tribe. It is a native of Mount Atlas and other hills on the coast of Africa. The wood of the tree was very valuable, and was used for inlaid tables, and it is noticed by Greek and Roman writers. It is still valued in Algiers. The wood is close-grained and fragrant. It yields the resin of the sandarach or pounce, used for scattering over manuscripts.

Thyine is from the Greek θύειν, to sacrifice.

The Arabs call the tree *alarz*, a word similar to the name *alerce* which is applied to the tree.

In 1796 Desfontaines found the thyine-tree on Mount

Atlas, where it had been seen previously by Pliny (*Natural History*, XIII. xxix. 15). Theophrastus speaks of the imperishable nature of the wood. It was used for building temples. The roof of the celebrated mosque, now the Cathedral of Cordova in Spain, built in the ninth century, was made of this wood.

It was used as fuel in sacrifices to heathen gods and to idols, and afterwards idols were made of the wood (Isaiah xliv. 16, 17). It is mentioned by Homer as acceptable in its perfume to the gods of Greece. According to Pliny, it was used sometimes for emperors' banqueting-tables, under the name of citron-wood. The wood is beautifully grained, and looks like fine mahogany. Cicero is said to have possessed a table made of this wood valued at about £12,000. The largest table of this wood belonged to Ptolemy, a king of Mauritania (the present Morocco).

But the thyine-tree, with the cedars of Lebanon, has been disappearing from the forests. The merchants no longer visit the marts of Babylon, of Greece, of Rome. No fires have been kindled upon the altar-stones of Baal, of Jupiter, and of Ashtaroth for ages. Many of those sacred altars have been brought down from their heights to furnish material for the cabins of the peasant and the hut and sheepcot of the Arab. Great changes, during centuries, have at last brought down the power of Babylon to the dust; and it is vanishing from off the earth. "The merchants of the earth have ceased to weep over her." "No man buyeth their merchandise any more," and "the fruits that their souls lusted after are departed;" "the merchandise of gold, and silver, and precious stones, and of pearls, and fine linen, and purple, and silk, and scarlet," as well as "all thyine wood." (Rev. xviii. 11–20.)

How completely has the vision of its beauty faded! The screech-owl and the bittern sit and mourn where songs once echoed from fragrant terraces, and where princes and nobles revelled over luxuries in the splendid halls of Babylon. (Isa. xiii. 19–22.)

## APPLE-TREE.

(See *Tappûach*. *Pyrus malus, Linn.*)

## TAPPÛACH.

(*Translated Apple-tree, and Apples.*)

"A word fitly spoken is like apples of gold [golden citrons] in pictures [baskets] of silver."—PROV. xxv. 11.

THE Hebrew word *tappûach* occurs in Proverbs xxv. 11; Song of Solomon ii. 3, 5, vii. 8, viii. 5; and in Joel i. 12. There have been great differences of opinion respecting the correct translation of this word. Rosenmüller and others render it "quince," while Royle renders it "citron," and says that its rich yellow colour ("citrons of gold," or "golden citrons"), its fragrant odour ("smell like citrons"), and the handsome appearance of the tree, whether in flower or in fruit, are particularly suited to all the passages of Scripture in which the word *tappûach* occurs. The Jews use the citron fruit at the present day at the Feast of Tabernacles. This is done from the idea that the word *etz'hadar*, translated "boughs of goodly trees" in Leviticus xxiii. 40, means branches of the citron-tree, which are thus associated with palm leaves, branches of thick trees (*etz'aboth*), and willows, in the Feast of Tabernacles.

The marginal reading of "boughs" is "fruit," and hence some think that citrons were the "goodly fruit" used along with leaves of palms. This view is taken by many of the rabbis and by Josephus. The citron is a native of Media, and is now very common in Palestine. The flowering branches are regularly used in the services of the synagogue on account of their sweet odour.

Tristram thinks that the word *tappûach* very probably applies to the apricot, which, though not a native of Palestine, was early introduced from Armenia, and now flourishes in Judæa. He says: "Many a time have we pitched our tents in its shade, and spread our carpets secure from the rays of the sun. 'I sat down under his shadow with great delight, and his fruit was sweet to my taste.' 'The smell of thy nose [shall be] like tappuach.' There can scarcely be a more deliciously perfumed fruit than the apricot; and what fruit can better fit the epithet of Solomon, 'Apples of gold in pictures of silver' than this golden fruit, as its branches bend under the weight in their setting of bright yet pale foliage?"

The citron is the produce of *Citrus medica* (Linn.), and belongs to the natural order Aurantiaceæ, or the Orange family. The apricot is *Prunus armeniaca* (Linn.), and belongs to the natural order Rosaceæ, sub-order Drupiferæ. The quince is *Pyrus cydonia*, and belongs to the natural order Rosaceæ, sub-order Pomaceæ.

## THORNS AND BRIERS AND BRAMBLES.

"All the land shall become briers and thorns."—ISA. vii. 24.

THE Hebrew words *atad, koz, chedek, choach, naazuz, shait, shamir, sillon, sirim, sirpad, zinnim*, and eight others, have been translated variously "thorns," "briers," and "brambles" in the Old Testament; and the word *akantha* is the "thorn" of the New Testament. It is impossible to say whether or not a particular species of plant was intended by each of these terms. Most of them apply generally to thorny plants, of which there are many in Palestine at the present day. Commentators mention among the thorny plants of the Holy Land species of *Zizyphus*, such as *Zizyphus spina-Christi*, also *Paliurus aculeatus, Acanthus spinosus, Ononis spinosa, Solanum spinosum, Tribulus terrestris, Lycium europœum*, and species of *Rhamnus, Centaurea*, and *Astragalus*.

Since man's fall, thorns of all kinds have come up on the ground, which was cursed (Gen. iii. 18); and God in chastening Israel often refers to the curse of thorns Thus Isaiah says, "Upon the land of my people shall come up thorns and briers" (xxxii. 13); and Hosea prophesies that "the thorn and the thistle shall come up on their altars" (x. 8).

The common bramble occurs in many parts of Palestine.

## LÔT, OR LADANUM.

(*Cistus creticus, Linn.*)

THE word *lôt* occurs twice in the Book of Genesis: "And they sat down to eat bread: and they lifted up their eyes and looked, and, behold, a company of Ishmeelites came from Gilead with their camels bearing spicery

LÔT, OR LADANUM.

and balm and myrrh, going to carry it down to Egypt" (Gen. xxxvii. 25). "And their father Israel said unto them, If it must be so now, do this; take of the best fruits in the land in your vessels, and carry down the man a present, a little balm, and a little honey, spices, and myrrh, nuts, and almonds" (Gen. xliii. 11). In both these cases the word *lôt* is translated "myrrh." This is a mistake, as myrrh is not a product of Gilead or of any part of Palestine. The substance referred to seems to be ladanum, obtained from a species of cistus, and more especially *Cistus creticus*. The plant belongs to the natural order Cistaceæ, the Rock-rose family. It has a large rose-coloured flower. The gummy matter which exudes from the plant is collected and used for its stimulating qualities.

*Cistus villosus* (Linn.) and *Cistus salviæfolius* (Linn.) are by some considered to be the myrrh referred to in Genesis xxxvii. 25.

## STACTE.

### (Hebrew, *Nataf*.)

"Take unto thee sweet spices, stacte, and onycha, and galbanum."
Ex. xxx. 34.

THE Hebrew word *nataf* corresponds with the Greek word *stacte*, which is mentioned as one of the ingredients of the holy incense which Moses was directed to prepare. It is supposed to be the substance called storax, which is procured from the *Styrax officinale* of Linnæus. It is a beautiful shrub, belonging to the natural order Styracaceæ, the Storax family. It is a native of Greece, Asia Minor, and Spain. Its flowers are white and have an orange odour, and the under side of its leaves is covered with a white down. It yields a fragrant resin which contains benzoic acid, and is used as a pectoral remedy. This plant abounds on the lower hills of Palestine. Tristram says that nothing can be more beautiful than the appearance of the storax in March, when covered with a sheet of white blossom, wafting its perfume through the dells of Carmel and Galilee, where it is the predominant shrub, and contrasts

STACTE.

beautifully with the deep red of the Judas-tree (*Cercis siliquastrum*) growing in the same localities. (See also remarks on *Poplar*.)

## PINE-TREE.

*(Hebrew, Tidhar.)*

"The glory of Lebanon shall come unto thee, the fir tree, the pine tree, and the box together."—Isa. lx. 13.

THE Hebrew word *tidhar* has been translated "pine-tree" in the passage above quoted, and also in Isaiah xli. 19.

It must refer to some companion tree on Lebanon; but commentators are unable to determine exactly the tree which is meant. Tristram thinks that the weight of evidence is in favour of the elm, a species of which—*Ulmus campestris* (?)—grows on Lebanon. We have already referred to "pine branches," mentioned in Nehemiah viii. 15. The Hebrew word thus translated is *etz 'shamen*, which more properly signifies the oil-tree or oleaster. (See *Oil-tree*.)

# ANISE OR DILL.

*(Peucedanum graveolens, Benth. and Hook.; Greek, Anethon.)*

---

"Ye tithe mint and anise."—MATT. xxiii. 23.

THE word *anethum* occurs in Matthew xxiii. 23, and has been translated "anise." The plant, however, referred to in this passage seems to be that known by the name of "dill," *Peucedanum graveolens* of botanists. This plant belongs to the natural order Umbelliferæ. The common dill is a herbaceous biennial plant, which is a native of the south of Europe and Egypt, and is also found near Astracan, Buenos Ayres, and at the Cape of Good Hope. The name is derived from the old Norse word to dill or soothe, referring to its carminative qualities in allaying gripes. It is one of the garden plants of which the Pharisees were in the habit of paying tithes. The plant is aromatic. It resembles fennel, and has finely divided leaves, which are used in pickles and soups. Pliny mentions it as a condiment (xix. 61, xx. 75). It is used also medicinally as a carminative in the form of distilled water of dill. The fruit yields a pale yellow oil, having a pungent odour, and an acrid, sweetish taste.

The true anise, *Pimpinella anisum*, has similar properties, and is also cultivated in Europe. The tithe of dill paid most punctually by the Pharisees in this and in other instances, was in conformity with the letter of

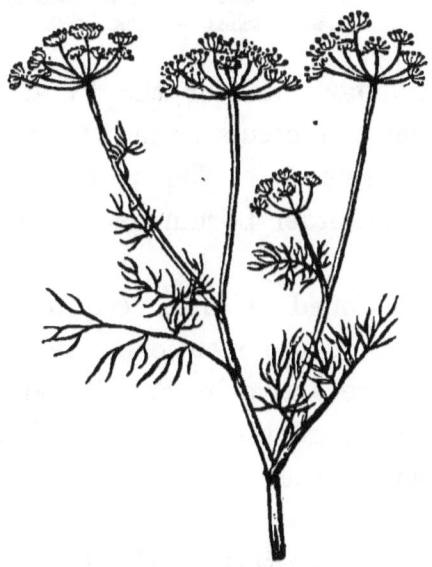

ANISE —(*Peucedanum graveolens.*)

the law, but they neglected more important matters. These they ought to have done, and not leave the other undone (Luke xi. 42).

# BEANS.

*(Vicia faba, Linn.; Hebrew, Pol; Greek, Cyamos: Latin, Faba vulgaris.)*

---

"Take thou also unto thee wheat, and barley, and beans, and lentiles, and millet, and fitches."—EZEK. iv. 9.

HE common bean is the *Vicia faba* of Linnæus, and occurs under the name of *pol* in the Hebrew. The plant belongs to the natural order Leguminosæ, and the sub-order Papilionaceæ.

Beans are mentioned twice in Scripture: in 2 Samuel xvii. 28, where they are mentioned among the articles of food brought to David by Barzillai and others; and in Ezekiel iv. 9, where the prophet is directed to take beans along with other vegetable products to make bread.

Beans ripen in Palestine at the time of the wheat harvest. Their flowers give off a fragrant perfume, and they are remarkable for the dark brown, almost black colour, seen on two of their petals.

Beans are mentioned by Theophrastus, the Greek philosopher and naturalist. The author of the "Scripture Garden Walk" says: "The meal of the bean was thought to stupify the senses and disturb the rest; certain characters also indicating heaviness and death were thought to be seen in its flowers. The Roman

BEANS.—(*Vicia faba.*)

tables were furnished with it at funerals and obsequies of the dead."

Beans are used as food by the poor, who knead the meal made from this with flour to make coarse bread.

# SWEET CANE.

(*Andropogon calamus-aromaticus*, Royle.)

---

"The sweet cane from a far country."—JER. vi. 20.

THE Hebrew words *kaneh-bosem* and *kaneh-hattob*, meaning reed of fragrance and fragrant reed, are translated "sweet cane" in our version of the Bible. The word *kaneh* is applied generally to a reed, and seems to be equivalent to the Latin *canna*, as well as to the Greek *calamos*, whence the name "culm" applied to stems of grasses. This sweet cane (Isa. xliii. 24) was an aromatic reed-like plant, remarkable for its fragrance, and imported from a far country (Jer. vi. 20). It is called "calamus" in Ezek. xxvii. 19, and "sweet calamus" in Ex. xxx. 23, and was used in compounding the holy ointment. Solomon mentions "calamus [roosa of India] and cinnamon, with all trees of frankincense" (Song of Sol. iv. 14). In the flourishing days of Tyrus, her merchants imported calamus. After examining the statements of Dioscorides and other ancient authors, Royle concluded that the sweet cane was an aromatic grass which he has called *Andropogon calamus-aromaticus*. It is a native of

India, where it is used in ointments and frankincense. It yields a fragrant oil called kuskuss, or roosa oil, or grass-oil. Royle states that the plant is found in Central India, that it extends as far north as Delhi, and as far south as between the Godavery and Nagpore, where it is called spear-grass. Another species, *A. schoenanthus*, is the lemon-grass or ginger-grass, which

SWEET CANE.—(*Andropogon calamus-aromaticus*)

some think to be the sweet cane of Scripture. It yields a fragrant oil. *A. citronum* supplies the perfume called citronelle; while *A. muricatus* furnishes kum-kus oil, which is used as medicine in India. All these canes are thus more or less sweet as regards their fragrance. They belong to the natural order Gramineæ (Grasses).

# CORIANDER.

*(Coriandrum sativum, Linn.)*

---

"And the manna was as coriander seed."—Num. xi. 7.

THE Hebrew word *gad* occurs in two passages of Scripture,—Ex. xvi. 31, and Num. xi. 7,—and has been translated "coriander." In both places the word is used to describe the manna, which was white and round like *gad*. There seems to be good reason for believing that the translation of the word is correctly given, and that the round fruit of the coriander is referred to. *Coriandrum sativum* is an annual plant, belonging to the natural order Umbelliferæ. The plant is about two feet high; its flowers are small and white, and are produced in umbels; and the fruit (often erroneously called seed) consists of two hemispherical carpels, which are so combined as to form a little ball or globe of the size of a pepper-corn. Each of these balls contains two seeds. The plant is very common in the south of Europe, and it grows also in India and other Eastern countries. It is cultivated in Britain on account of its seeds and fruit, which are used by confectioners, druggists, and distillers. About fifteen tons of the fruit are annually imported from Germany. The leaves are used as a salad. The Greek name of the plant is *korion* or *koriannon*, whence the English name coriander. The fruit has an aromatic taste and smell, and yields by distillation a volatile oil, to which its properties are due.

CORIANDER.—(*Coriandrum sativum.*)

*Man*, or *manna*, rendered "manna" in the Bible, is the name applied to the food with which God fed the Israelites in the desert (Deut. viii. 3; Neh. ix. 20; Ps. lxxviii. 24; John vi. 31, 49, 58; Heb. ix. 4). The food was miraculously brought to the encampment. We know nothing about it except that it was in small round grains like coriander seed or fruit (Ex. xvi. 31; Num. xi. 7), and that it tasted like wafers and honey. Some authors have supposed that there are similar substances produced by plants at this day in the East, and which are now called manna. Among these are reckoned exudations from *Tamarix gallica*, French tamarisk, and *Alhagi maurorum*, or camels' thorn. Royle remarks, " None of these mannas explain—nor can it be expected that they should explain—the miracle of Scripture by which abundance of manna is stated to have been produced for millions in a country where hundreds cannot now obtain subsistence."

# CORN.

"He will also bless thy corn and thy wine."—DEUT. vii. 13.

HERE are various Hebrew words translated "corn." Among these are *bar, dagan, kamah,* and *shibboleth,* the latter meaning an ear of

WHEAT, BARLEY, AND RYE.

corn (Gen. xli. 5; Ruth ii. 2). The ordinary cultivated grains in Palestine were wheat, barley, and "spelt" (translated rye).

# CUMMIN.

*(Cuminum cyminum, Linn.)*

"Ye tithe mint, and anise, and cummin."—MATT. xxiii. 23.

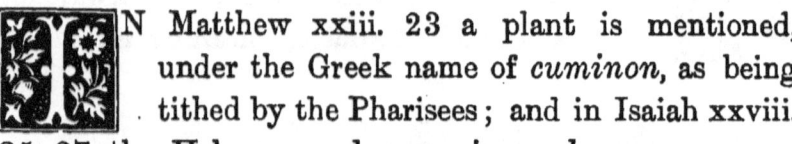N Matthew xxiii. 23 a plant is mentioned, under the Greek name of *cuminon*, as being tithed by the Pharisees; and in Isaiah xxviii. 25, 27 the Hebrew word *cummin* or *kammon* occurs. Both seem to refer to the plant called cummin at the present day, the *Cuminum cyminum* of botanists. The plant belongs to the natural order Umbelliferæ. It is an annual plant, bearing whitish or reddish flowers, and yielding an aromatic fruit. What are commonly called cummin-seeds are really single-seeded fruits. They contain a fragrant volatile oil. The plant is said to be a native of Upper Egypt and Ethiopia, but it is cultivated in Eastern countries and also in the south of Europe, and its fruit is used as a medicine and a condiment. Britain receives its supply of cummin chiefly from Malta and Sicily.

Isaiah mentions the cultivation of cummin in ancient times: "When he [the ploughman] hath made plain the face thereof, doth he not cast abroad the fitches, and scatter the cummin?" (Isa. xxviii. 25.) And he alludes to the mode in which the fruit was reaped when he says: "For the fitches are not threshed with a threshing instrument, neither is a cart wheel turned upon the cummin; but the fitches are beaten out with a staff, and the cummin with a rod" (Isa. xxviii. 27). This mode

CUMMIN —(*Cuminum cyminum.*)

of preparation is required in the case of cummin, the fruit of which is easily separated by a light shake; but if bruised by a wheel, it would be injured, inasmuch as the oil, to which it owes its properties, would be pressed out. The scribes and Pharisees were condemned by our Lord because, while they paid tithe of cummin, they neglected the weightier matters of the law, —judgment, mercy, and faith. The formal offering was made, but there was no life-giving spirit.

It is said that in nations where the rite of circumcision was practised, bruised cummin fruit mixed with wine was used as a styptic after the operation.

In 1870 there were imported into Britain 2,385 cwts. of cummin fruits.

# FITCHES.

(*Nigella sativa, Linn.*)

---

"The fitches are beaten out with a staff."—Isa. xxviii. 27.

THE words *ketzach, kezach*, and *ketzah*, or *quetsah*, occur in the books of the prophets Isaiah and Ezekiel, and have been translated "fitches." These fitches, or vetches, appear to have been sown like other grain, and Isaiah alludes to them in speaking of the different occupations of the husbandman. Thus he says: "Doth the plowman plow all day to sow? doth he open and break the clods of his ground? When he hath made plain the face thereof, doth he not cast abroad the fitches, and scatter the cummin, and cast in the principal wheat [that is, the wheat in the principal place] and the appointed barley [that is, barley in the appointed place] and the rye in their place?" (Isa. xxviii. 24, 25.) Reference is here made to the appointed season of husbandry and the variety of the operations carried on; and thus the Lord calls the attention of his people to his different modes of dealing with them.

# FITCHES.

In Ezekiel iv. 9 there occurs the word *cussemeth*, which has been translated "fitches" in place of spelt.

There is some difficulty in ascertaining what plant is meant by the term "fitches." Some have referred it to the common vetch (*Vicia sativa*); but this does not seem to be correct.

In the Septuagint the name *melanthia* is given,

FITCHES.—(*Nigella sativa.*)

derived in part from the word *melas*, "black." The name is considered as referring to a plant with black seed; and after careful comparison of names, Royle and other authorities have concluded that the plant is the *Nigella sativa* of botanists. The vulgar name of *kezach* is stated to be *nielle*,—that is, *nigelle*. This plant is called *melanospermum,* or "black seed," by the Greeks; and its Arabic name means the same thing, while the Latin, *nigella,* also indicates blackness. By old Latin authors the plant was called *git* or *gith,* and it is referred to by Pliny (lib. xx. cap. 17). The plant is commonly cultivated in the East. Its seeds have aromatic qualities, and were used like pepper as a condiment with food. Hence they are mentioned with other aromatic and carminative plants, such as cummin and anise. They are sometimes called black cummin. Dioscorides and Pliny refer to their use in bread.

The seeds are easily beaten out of the seed-vessel; and allusion is made to this in Isaiah xxviii. 27, where the prophet says, "The fitches are beaten out with a staff;" whereas the cummin required a rod.

The genus *Nigella* receives the English name of fennel-flower, from its leaves resembling those of fennel. It belongs to the natural order Ranunculaceæ or Crowfoots, and the Hellebore section of that family. There are ten known species. They are erect, annual, herbaceous plants, found in the Mediterranean region as well as in Western Asia. Their flowers are solitary at the tops of the stems or branches, and they are of a whitish blue or yellow colour. They have a coloured calyx, small petals, acrid aromatic seeds, and finely-cut leaves. Their seed-vessel consists of numerous carpels, more or less united together, and opening on the inner side so as to scatter the seeds.

# FLAX.

(*Linum usitatissimum, Linn.*)

---

" Smoking flax shall he not quench."—Isa. xlii. 3; Matt. xii. 20.

THE Hebrew word *pishtah* has been proved to mean the flax or lint plant. It is the *Linum usitatissimum* of botanists, and the *linon* of the Greeks. The plant belongs to the natural order Linaceæ, the Flax family. The species are herbs or under-shrubs, with narrow undivided leaves, and blue, red, or white flowers, arranged in racemes or clusters. The number *five* prevails in the genus. Thus the plants have five sepals, five petals, five perfect and five abortive stamens, and five or ten divisions of their seed-vessel. There are about eighty known species found in the temperate and warm or intertropical regions of the Old and New Worlds, and a few in the tropical parts of South America.

The cultivated flax plant has a blue flower. It yields fibres which are used in the manufacture of linen. Its seeds yield oil; and the substance left, after the oil has been expressed, is the oil-cake which is given as food to cattle. Frequent references are made in the Bible to flax and linen. In Egypt the flax plant was extensively cultivated, and employed for the manufacture of linen. The cloth made from it was used to wrap the mummies. By examining the mummy cloth under the micro-

FLAX.—(*Linum usitatissimum.*)

scope, we ascertain that it was formed from flax, and not from cotton. On various Egyptian monuments the plant and the preparation of its fibres are represented. The usual mode in which flax is prepared is by steeping it in water, allowing all the softer parts to be removed, and retaining the fibrous portion. The process is tedious, and is often accompanied with injury to the fibres. Of late years great improvements have taken place in the manufacture.

The first allusion to the flax plant in the Bible is when the plague of hail was sent by God as a judgment on the Egyptians: "And the flax and the barley was smitten: for the barley was in the ear, and the flax was bolled" (Ex. ix. 31). The period of the year when the plague was sent was spring, probably about April, a time when hail-storms were very uncommon. It would appear, therefore, that, as is the custom in India now-a-days, the flax and the barley were sown in the months of September and October, and the reaping took place in the early part of summer. The flax being "bolled" means that the flower-buds were formed. Some have translated the passage, "the flax was in blossom." Whichever translation is taken, it is clear that the flax was far advanced, so as to be injured by the hail. God showed his power and sovereignty by destroying one of the sources whence the Egyptians derived articles of comfort and luxury. That flax was cultivated in Palestine is shown in Joshua ii. 6, where it is stated that the faithful Rahab used flax to hide the spies sent by Joshua to examine Jericho: "But she had brought them up to the roof of the house, and hid them with the stalks of flax, which she had laid in order upon the roof." In the history of Samson, also (Judges xv. 14), reference is made to flax as being well known. (See also Hosea ii. 5, 9.) The spinning of flax by the hand with the spindle and distaff is alluded to in Proverbs xxxi. 13, 19, where it is said of the virtuous woman, "She seeketh wool and flax, and worketh willingly with her hands...She layeth her hands to the spindle, and her hands hold the distaff."

This mode of preparing yarn is portrayed on the marbles of Athens and Rome, and is still practised in some countries. The working of fine flax and linen was an important manufacture, and the destruction of those employed in it is mentioned by Isaiah as one of the awful judgments to be inflicted on Egypt: "Moreover, they that work in fine flax, and they that weave networks [or white works], shall be confounded" (Isa. xix. 9).

Flax seems to have been put to various uses, as in the preparation of linen clothing, curtains, ephods, girdles, mitres, bonnets, ropes, and wicks. In many passages in Exodus, Leviticus, Deuteronomy, and Chronicles allusions are made to the use of linen, and fine linen, in the formation of the priests' garments and of the hangings of the tabernacle. Samuel ministered before the Lord with a linen ephod (1 Sam. ii. 18). David danced before the ark, girded with a linen ephod (2 Sam. vi. 14). Jeremiah was told by the Lord to get a linen girdle, and put it upon his loins (Jer. xiii. 1). Solomon had linen yarn brought out of Egypt (1 Kings x. 28; 2 Chron. i. 16). Ezekiel speaks of a cord or measuring-line of flax (Ezek. xl. 3). Hosea also refers to flax as used for making garments (Hos. ii. 5, 9).

The words translated "linen" and "fine linen" in these passages are *shesh* or *sheshi*, *bad*, and *butz*. Some suppose that *shesh* refers to hemp. The word resembles the Arabic name *haschesch*, which is applied to hemp. There is much difficulty in determining the sources whence the linen of Scripture was derived. (See *Hemp*.)

The use of flax for wicks is referred to by Isaiah. When describing the tenderness and love of the Saviour, he says, "A bruised reed shall he not break, and the smoking flax shall he not quench" (Isa. xlii. 3). This passage is also quoted in Matthew xii. 20, the only place in the New Testament where the word flax occurs.

In the New Testament, linen is mentioned on several occasions. The Greek word in these passages is *byssus*. It was in linen that the body of Christ was wrapped (Matt. xxvii. 59; Mark xv. 46, John xix. 40); and the linen clothes were seen by the disciples when they visited the tomb (Luke xxiv. 12; John xx. 5–7).

Fine linen constituted the clothing of the rich and great in former times. Pharaoh arrayed Joseph in a vesture of this kind (Gen. xli. 42). Mordecai went out from the presence of the king with a garment of fine linen (*butz* or *buz*) and purple (Esther viii. 15).

Fine linen (*shesh*) is mentioned by Isaiah and Ezekiel as one of the luxuries of Judah, Jerusalem, and Tyre (Isa. iii. 23; Ezek. xvi. 10, 13, xxvii. 7, 16). The rich man was clothed in purple and fine linen (Luke xvi. 19). Fine linen is recorded among the costly merchandise of mystic Babylon, over the loss of which the merchants of the earth shall weep and mourn (Rev. xviii. 12, 16). In the last two passages the word used for fine linen is *byssus*.

# FRANKINCENSE.

(*Boswellia thurifera, Colebr.*)

---

"Calamus and cinnamon, with all trees of frankincense."—SONG OF SOL. iv. 14.

RANKINCENSE is often mentioned in the Bible. It was furnished by many trees, especially by species of Boswellia. These have been well described by Dr. Birdwood in the "Transactions of the Linnæan Society," vol. xxvii., p. 111. The Hebrew word for incense is *lebonah*. In Exodus xxx. 34, pure frankincense is noticed as one of the ingredients of the sweet spices of which the pure and holy perfume was made after the art of the apothecary, which was offered every morning and evening on the altar of incense. In the Song of Solomon allusion is made to frankincense (iii. 6, iv. 14). The word "incense" meaning frankincense is also often used in the Old Testament: "To what purpose cometh there to me incense [frankincense] from Sheba [Arabia], and the sweet cane [*roosa*] from a far country?" (Jer. vi. 20.) "I have not caused thee to serve with an offering, nor

wearied thee with incense [frankincense]. Thou hast brought me no sweet cane [*roosa*] with money" (Isa. xliii. 23, 24). "The multitude of camels shall cover thee, the dromedaries of Midian and Ephah; all they from Sheba [Arabia] shall come: they shall bring gold and incense" [frankincense] (Isa. lx. 6). The wise men from the east when they came to worship Jesus "presented unto him gifts; gold, frankincense, and myrrh" (Matt. ii. 11). Frankincense is produced chiefly in Arabia-Felix and the Soumali country. There is a kind of frankincense called *olibanum*. It is the produce of *Boswellia carterii* and *Boswellia bhau-dajiana* of Birdwood. The plants belong to the natural order Burseraceæ, the Myrrh family.

FRANKINCENSE.
(*Boswellia thurifera.*)

# GALBANUM.

*(Polylophium officinale, Benth. and Hook.)*

---

"Sweet spices, stacte, and onycha, and galbanum."—Ex. xxx. 34.

THE Hebrew word *chalbaneh,* or *chelbena,* occurs only once in the Bible, and it has been translated "galbanum." We know that it was used along with other spices in forming "an oil of holy ointment, an ointment compound after the art of the apothecary...an holy anointing oil." It is by no means clearly ascertained what the substance was. The galbanum of the present day is a fetid gum-resin procured from an umbelliferous plant, and is imported from India and the Levant. The substance occurs in irregular pieces, about the size of a pea, which are usually agglutinated into masses of a greenish-yellow colour, having a strong disagreeable odour, and an acrid bitter taste. Some authors name the plant *Galbanum officinale.* The gum-resin exudes from the plant, and is collected in tears. Many umbel-bearing plants yield gum-resins

GALBANUM.—(*Polylophium officinale.*)

of a similar character, which are used by Eastern nations as condiments, although not very palatable to Europeans.

# WILD GOURD.

*(Citrullus colocynthis, Schrad.)*

"And gathered thereof wild gourds his lap full."—2 KINGS iv. 39.

THE Hebrew word *pakyoth*, translated "wild gourds," occurs in 2 Kings iv. 39, where it is stated that when Elisha came to Gilgal he told his servant to set on the great pot, and seethe pottage for the sons of the prophets; and that "one went out into the field to gather herbs, and found a wild vine, and gathered thereof wild gourds his lap-full, and came and shred them into the pot of pottage;" and that when they were eating it they cried out that there was "death in the pot."

It is obvious, from the narrative, that the person who went to gather the herbs had made a mistake in regard to them, and had taken some nauseous and poisonous fruit instead of what was wholesome. From a careful examination of the Hebrew word and of the Arabic version of it, commentators are disposed to think that the wild gourd was the fruit of the colocynth plant, the coloquintida or bitter apple, *Cucumis* or *Citrullus colo-*

*cynthis* of botanists. It belongs to the natural order Cucurbitaceæ, the Cucumber family. The fruit of it is used medicinally as an active purgative. It resembles a wild vine, in its trailing mode of growth and in its tendrils. Its fruit is round, orange-coloured, and tempting in appearance; but the pulp of it is very bitter, and hurtful when taken even in small quantities. In all these respects, then, it would agree with the facts as given in the Scriptures. It was probably mistaken for some of the other species of cucumber or gourd which have eatable fruit. As the plant is not generally diffused over Palestine, it might be unknown to the sons of the prophets. In the tribe of plants to which the cucumber, melon, gourd, and vegetable marrow belong,

COLOCYNTH PLANT.—(*Citrullus colocynthus*)

there are several bitter, acrid, and even poisonous species. Besides the colocynth, there is another, called the squirting cucumber, which acts as a poison. It receives its name from the fact that when ripe the seeds are squirted out from the interior of the fruit with great force, at the point where the stalk is attached. Some authors think that, as the Hebrew word is derived from the verb "to burst," the plant may be this cucumber. In the young state, the fruit is very like the young cucumbers called gherkins. The fruit is a drastic purgative, and acts as a poison. Both of these plants are found in Palestine;

Wild Cucumber squirting Seed.

and the colocynth, in particular, trails along the ground in a luxuriant manner there. It abounds in the desert parts of Syria, Arabia, and Persia, and on the banks of the Euphrates and Tigris. Many miles of country are sometimes covered with this plant, which bears a prodigious number of gourds. The fruit is imported into this country from Smyrna, Trieste, France, and Spain.

There is also another species of cucumber having similar properties, and called at the present day the prophets' or globe cucumber—*Cucumis prophetarum*—probably from an allusion to the statements in the Bible relative to wild gourds. The fruit is small, being not larger than a cherry. There is a Hebrew word, *pekaim*, translated "knops," which occurs in 1 Kings vi. 18, and vii. 24, and which seems to be derived from *pakyoth*. These knops were of a rounded form, and were probably made in imitation of the fruit of the wild gourd.

# HEMP.

### (*Cannabis sativa, Linn.*)

---

"Thy raiment was of fine linen."—EZEK. xvi. 13.

HE Hebrew word *shesh* or *sheshi*, translated "fine linen," occurs, according to Royle, twenty-eight times in Exodus, once in Genesis, once in Proverbs, and three times in Ezekiel. This fine linen was spun by women, as mentioned in Exodus xxxv. 25, where it is said, "All the women that were wise hearted did spin with their hands, and brought that which they had spun, both of blue, and of purple, and of scarlet, and of fine linen." Ezekiel says of Tyrus, "Fine linen with broidered work from Egypt was that which thou spreadest forth to be thy sail" (Ezek. xxvii. 7). The material of which this fine linen was wrought is considered by many to have been the produce of the hemp plant. This is rendered probable also by the similarity between *shesh* and the Arabic word *haschesch*, which is applied to hemp. Hemp consists of the fibres of *Cannabis sativa*, a plant belonging to the natural order Urticaceæ or nettleworts. It is a native of Persia, and

is now extensively cultivated in Europe as well as in India. The variety cultivated in India is sometimes called *Cannabis indica,* and is remarkable for its narcotic qualities. The dried flowering tops of the female plant from which the resin has been removed are used to form a medicinal extract and tincture. The

Hemp.—(*Cannabis sativa.*)

resinous matter covering the leaves is called *churrus;* and the names *bhang, gunjah,* and *haschesch,* are given to the dried plant in different states. It seems likely that the hemp plant was cultivated in Egypt in ancient times as well as the flax plant; but accurate information on the subject is still wanting. The Hebrew word *bad* is also translated "linen." Thus it occurs in Exodus xxxix. 28, where it is said that they made for Aaroh and his sons "a mitre of fine linen, and goodly bonnets of fine linen, and linen breeches of fine twined linen." The Hebrew word *butz* or *buz* is also translated "fine linen" and "white linen," as in 1 Chronicles iv. 21; Esther i. 6; Ezekiel xxvii. 16, etc. In the New Testament the Greek word *byssus* is translated "fine linen," as in Luke xvi. 19; Rev. xviii. 12, 16, and xix. 8, 14.

(See also *Flax.*)

# SAFFRON.

### (*Crocus sativus, Linn.*)

---

"Thy plants are...spikenard and saffron."—SONG OF SOL. iv. 13, 14.

THE Hebrew word *karcom* or *carcom* occurs in the Song of Solomon iv. 14, where it is translated "saffron." It is there mentioned along with other fragrant substances and spices, as spikenard, calamus, cinnamon, frankincense, myrrh, and aloes. In the Greek it is rendered by the word *krokos*, and the Arabic name is *safran*. The plant is the *Crocus sativus* of botanists. It belongs to the natural order Iridaceæ or Iris family. It is mentioned by ancient classical authors, and it has been cultivated from the earliest times in Asiatic countries. At the present day it is grown extensively in Persia and Kashmir. The saffron of commerce is imported from Spain, France, and Naples. It is a portion of the central part of the flower called the style or the stigma. This portion is removed and dried. It is of an orange-brown colour, and has a powerful aromatic odour. When rubbed on a moistened finger, it tinges it intensely orange-yellow. Cake saffron is formed by the stigmas being pressed together, and is imported from Persia into India. In Eastern countries saffron was highly esteemed as a kind of spice, which was used along with food. In India at the present day saffron from Kashmir is employed to colour and flavour native dishes. The cultivation of the saffron crocus was recommended by ancient writers as a means of attracting bees. The plant is cultivated in England, as at Saffron-Walden in Essex, which receives its name from that circumstance. Costly perfumes were made from the plant. Rosenmüller says that "not only saloons, theatres, and places which were to be filled with a pleasant fragrance, were strewed with this substance, but all sorts of vinous tinctures retaining the scent were made of it; and the perfume was poured into small fountains, which diffused a highly-esteemed odour. Even fruit and comfitures placed before guests, and the ornaments of the rooms, were spiced over with it. It was used for the same purposes as the modern pot-pourri." Saffron is used medicinally in the form of tincture and as an ingredient of aromatic powder.

The fragrance of this plant and others is used in Scripture to shadow forth the graces of the Christian as brought out by the Sun of Righteousness, and by the breathing of the Spirit, who blows upon the garden and makes the spices thereof flow out.

# LENTILES.

### (*Ervum lens, Linn.*)

---

"Jacob gave Esau...pottage of lentiles."—GEN. xxv. 34.

THE Hebrew word *adashim* occurs in several places in the Old Testament, and it has been translated "lentiles" or "lentils." These are the seeds of a kind of pulse called *Ervum lens* or *Lens esculenta* by botanists. The plant is the *phakos* of the Greek, the *addas* of the Arabic. It is an annual, and is the smallest of the cultivated plants of the pea-tribe. It flowers in May and ripens its fruit in July. The seeds contained in the pods are small and flattened, resembling a double-convex *lens* or magnifying glass; hence their name. The plant belongs to the natural order Leguminosæ, sub-order Papilionaceæ. It is cultivated in the south of Europe, Bombay, Egypt, and the Levant. Virgil in the "Georgics" speaks of the Pelusian lentile. It is sometimes used as fodder in England; and an attempt was made not long ago to raise it as pulse in a sheltered locality in Scotland. It is a weak plant, attaining a height of eighteen inches, supporting itself by tendrils which twine round other plants. Its leaves are compound, with usually eight pairs of leaflets in each, and have lanceolate, fringed stipules. The peduncles are usually two-flowered, and are about as long as the leaves. The flowers are purple and pea-like.

LENTILES.—(*Ervum lens.*)

The fruit is a short pod containing two or three seeds. The seeds supply nutritious food, and are employed for making pottage, which is of a yellowish hue or reddish colour. The red pottage which Jacob supplied to Esau, and for which the latter sold his birthright, was made of lentiles (Gen. xxv. 29–34).

Lentiles were cultivated like pease and beans, and we find in 2 Sam. xxiii. 11, an allusion to a field of lentiles which was protected from the Philistines by Shammah, one of David's mighty men. Lentiles are noticed among the provisions brought by Shobi, Machir, and Barzillai to David, when he was in the wilderness on account of the rebellion of Absalom (2 Sam. xvii. 28).

In times of scarcity lentiles were mixed with wheat, barley, millet, and fitches, in making bread (Ezek. iv. 9). In the southern parts of Egypt it appears that lentiles with a little barley formed almost the only bread used by the poorer classes. Some of the paintings on the tombs of the ancient Egyptians represent the cooking of lentiles and the preparation of pottage from them.

In some Roman Catholic countries lentiles are used as food during Lent; and some say that the name of the season is derived from this circumstance.

The Arabs have a tradition that the place where Esau sold his birthright is in Hebron, near the cave of Machpelah; and it is said that a college of dervises near the spot daily cook pottage of lentiles mixed with pot-herbs, for distribution among the poor.

# RUE.

(*Ruta graveolens, Linn.*)

---

"Ye tithe mint and rue."—LUKE xi. 42.

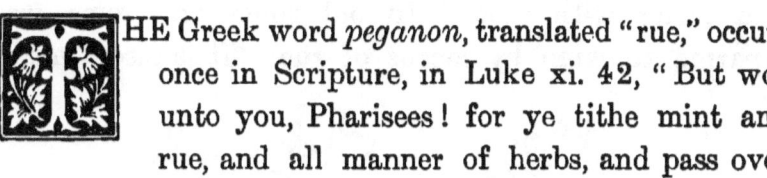HE Greek word *peganon*, translated "rue," occurs once in Scripture, in Luke xi. 42, "But woe unto you, Pharisees! for ye tithe mint and rue, and all manner of herbs, and pass over judgment and the love of God: these ought ye to have done, and not to leave the other undone." In the parallel passage, Matthew xxiii. 23, *anethon* or *dill* (translated "anise"), is named instead of rue. No doubt both were mentioned by our Lord, and each is recorded by a different evangelist. Both of these herbs were cultivated in Eastern gardens, as they are at the present day. Rue is a strong-scented plant (*Ruta graveolens* of botanists) which abounds in oil. The plant belongs to the natural order Rutaceæ. The plants belonging to the rue family are remarkable for the volatile oil which they yield. One of them, the dittany or fraxinella (*Dictamnus fraxinella*), is said to give out so much oily vapour in a warm, still evening, that the air around it

becomes inflammable.  Rue grows wild in the south of Europe and in Palestine.  Hasselquist mentions having seen it on Mount Tabor.  It is cultivated as a pot-herb, and more especially as a sort of spice or condiment to be used along with food.  In old times an aroma was imparted to wine by means of rue.  It is also a medi-

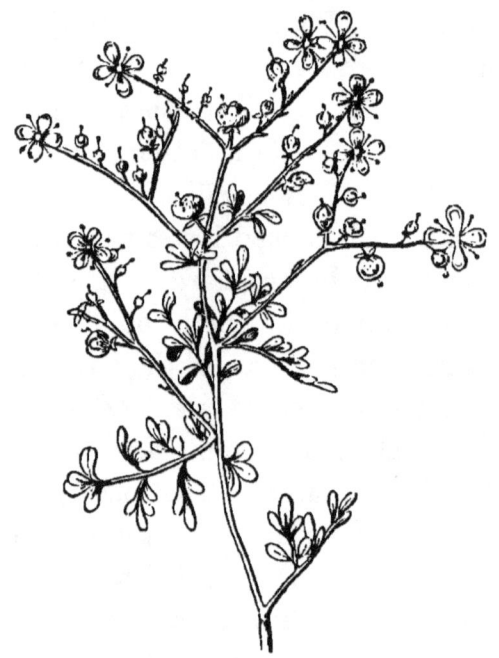

RUE.—(*Ruta graveolens.*)

cinal plant, and has been prescribed to allay spasms. Oil of rue is distilled in England from the fresh leaves and the unripe fruit.  It has a pale yellow colour, a disagreeable odour, and a bitter, acrid taste.  The tithing of it by the Pharisees calls for the same remarks that have already been made regarding mint, anise, and cummin.  They were very particular in regard to outward, legal observances, and even went beyond what was required; but, alas! they had not the spirit of the commandment in their hearts. They neglected weightier matters, judgment, and the love of God; and they brought upon themselves the condemnation of our Lord.

Rue was anciently called "herb of grace," and it is referred to under this name by Shakespeare :—

> "Here in this place
> I'll set a bank of rue, sour herb of grace."

From this we have the word *rue*, meaning repentance, which is needful to obtain God's grace. (Prior on *Popular Names of Plants.*)

# MINT.

*(Mentha sylvestris, Linn.)*

"Ye pay tithe of mint and anise and cummin."—MATT. xxiii. 23.

THE Greek word *heduosmon*, or *heduosmos*, which means "having a sweet smell," occurs in two passages of the New Testament, Matthew xxiii. 23, and Luke xi. 42, and has been translated "mint." It corresponds with the Latin *mentha*. The species of mint most common in Syria is that represented in the figure, and called by botanists *Mentha sylvestris*. It is often cultivated in gardens, and it is generally distributed over Europe, and reaches even to Kashmir. It is likewise found in Britain. The plant belongs to the natural order Labiatæ. It is an erect plant, with opposite, nearly sessile, ovate, lanceolate, and downy leaves, which are whitish below. The spikes of flowers are dense, and have a conically-cylindrical form. Another species is also common in Palestine, and is called field-mint (*Mentha arvensis*). The species of mint have all carminative qualities. They grow usually in damp places, and have reddish flowers arranged in spikes or whorls.

# MINT.

Mint was much used as a condiment in ancient times, from its aromatic qualities, in the same way as it is employed at the present day for a sauce to lamb. Pliny mentions it as highly esteemed. It was easily propagated, and its cultivation was attended with very little expense.

MINT.—(*Mentha sylvestris.*)

In Scripture it is noticed along with other sweet herbs, such as anise or dill, cummin and rue, which are commonly found in European gardens at the present day.

The giving of the tenth part to the Lord was enjoined on the Jews; and the Pharisees were very particular as to the letter, tithing even the smallest products of the garden: but they did it not in a right spirit; for they neglected the weightier matters of the law —judgment, mercy, and truth. These ought they to have done, and not to leave the others undone.

Lady Callcott, in her "Scripture Herbal," says: "I know not whether mint was originally one of the bitter herbs with which the Israelites ate the paschal lamb; but the use of it with roast lamb, particularly about Easter time, inclines me to suppose it was."

# ROSE.

*(Narcissus tazetta, Linn.)*

---

"I am the rose of Sharon, and the lily of the valleys."—SONG OF SOL. ii. 1.

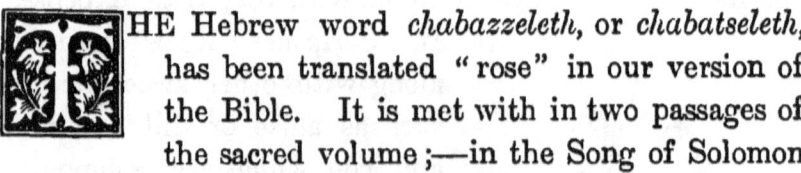HE Hebrew word *chabazzeleth*, or *chabatseleth*, has been translated "rose" in our version of the Bible. It is met with in two passages of the sacred volume;—in the Song of Solomon ii. 1, "I am the rose of Sharon, and the lily of the valleys;" and in Isaiah xxxv. 1, "The wilderness and the solitary place shall be glad for them; and the desert shall rejoice, and blossom as the rose." It would appear, however, from the researches of Celsius and other learned authors, that in place of the rose a bulbous plant is referred to—in all probability a species of narcissus. Royle considers the plant as probably *Narcissus tazetta*, the Polyanthus narcissus.

This plant belongs to the natural order Amaryllidaceæ, the Amaryllis family. Its white, fragrant flowers, are pushed forth in clusters from sheathing leaves, and it has a corona or crown in the centre of the flower. It is found in Palestine and in Syria, and it is highly esteemed both for its beauty and its fragrance. It is one of the plants which deck the meadows in spring with their blossoms. It seems to have adorned the level tract along the Mediterranean between Mount Carmel and Cæsarea, and which was known as the rich plain of Sharon. Hence the name "rose of Sharon." The fertility and richness of this plain are alluded to by Isaiah when he

Rose of Sharon.—(*Narcissus tazetta.*)

speaks of "the excellency of Carmel and Sharon" (Isa. xxxv. 2). Canon Tristram is disposed to think that the *Anemone coronaria* is the true rose of Sharon.

The plant is employed in Scripture to shadow forth Him "who offered himself a sacrifice to God for a sweet-smelling savour;" and to picture the blessedness of that time when the earth shall be full of the knowledge of the Lord.

In some of the apocryphal books we meet with the word which properly means *rose*, the *rhodon* of the Greeks.

Roses are highly prized in the East, and many wild species have been observed in Syria. The damask and hundred-leaved rose are cultivated extensively. What has been called the rose of Jericho is a species of cruciform plant, *Anastatica hierochuntica*, which, after flowering, dries up into a sort of ball.

# MILLET.

*(Panicum miliaceum, Linn.)*

"Take thou also unto thee wheat, and barley, and beans, and lentiles, and millet."—EZEK. iv. 9.

THE Hebrew word *dokhan*, or *dochan*, occurs in Ezekiel iv. 9, where the Lord says to the prophet, "Take thou also unto thee wheat, and barley, and beans, and lentiles, and millet, and fitches, and put them in one vessel, and make thee bread thereof." These are all plants which are used at the present day to furnish articles of food in Eastern countries. The millet is the produce of *Panicum miliaceum*. It is the *cenchros* of the Greeks. The grain is called *warree* in the East Indies. It belongs to the natural order Gramineæ or Grasses. Some suppose that *Panicum italicum* and *Sorghum vulgare*, the great millet or sowaree, may also be included in the Hebrew word. Both of these are grasses which furnish materials for bread. The common millet is imported from the Mediterranean into Britain. It is sometimes grown in

England to supply birds' seed. The plant has an erect stalk or culm from two to four feet high, with large leaves and a nodding cluster of fruit. In India and Persia at the present day it is extensively used for food, and it is often mixed with other grain to form bread.

MILLET.—(*Panicum miliaceum.*)

There is another grain resembling millet which is imported as an article of food. It is the *Andropogon sorghum*, durra, or dourra. It is said to be represented, along with other grains, on the ancient tombs of Egypt.

# TARES.

*(Lolium temulentum, Linn.)*

---

"His enemy came and sowed tares [darnel] among the wheat."
MATT. xiii. 25.

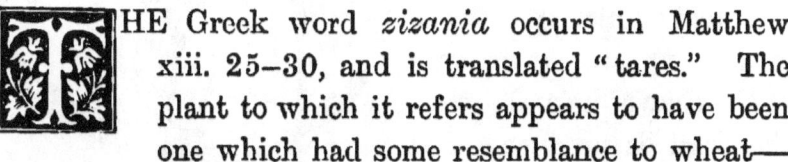HE Greek word *zizania* occurs in Matthew xiii. 25–30, and is translated "tares." The plant to which it refers appears to have been one which had some resemblance to wheat—at least in the blade—and hence totally unlike the plant called tares now-a-days, which is a kind of vetch. It is said, "But while men slept, his enemy came and sowed tares among the wheat, and went his way. But when the blade was sprung up, and brought forth fruit, then appeared the tares also." It is stated, also, that there was a difficulty in separating the one from the other, and a risk of pulling up the wheat as well as the tares in the attempt of the servants to get rid of the latter.

From a careful investigation of the matter, it has been supposed that the tare (*zizanion*) was the plant called darnel-grass (*Lolium temulentum*), which, while it has some resemblance to wheat, differs from it totally in quality. The darnel is a noxious grass, having narcotic qualities to a certain extent; and hence the necessity for rooting it out. An attempt to do so, especially in the early stages of growth, might be unsuccessful, from the great similarity between it and wheat. It is only at the time of harvest, when the fruit is produced, that the two crops can be accurately distinguished. The plant is the *ziwan* or *zawan* of the Arabs, the *infelix lolium* of Virgil (Georg. i. 154). Bad wheat sent from the Continent often contains darnel. Darnel is found in Palestine and Syria; and the grains of it when eaten are said to produce at the present day giddiness and stupefaction. Dr. Robinson says that among splendid fields of wheat near Kübrikhah are still found tares. They are like the wheat, and are not to be distinguished until the ear appears. The grain resembles wheat in form, but is smaller and dark. In Beirüt poultry are fed upon the

TARES OF PALESTINE.
(*Lolium temulentum.*)

grain, and it is kept for sale for that purpose. When not separated from the wheat, bread made from the flour often produces deleterious effects in persons who eat it.

Both wheat and darnel belong to the natural order Gramineæ or Grasses. They are easily distinguished when in flower, by the wheat having two glumes, and its florets having their edges next the rachis or common stalk; while in the darnel there is one glume. and the florets have their backs next the rachis.

The mode of gathering the harvest in Palestine resembles in some instances that mentioned in the parable. When the millet crop, for instance, is ripe, the reapers pull it up with their hands, and along with it the weeds that have grown up beside it, and then separate them.

The tares represent those false professors who are associated in this world with the wheat—that is, the true people of God. Both grow up together, and may at first seem alike, just as the wise and foolish virgins appeared to be. Man cannot distinguish accurately between the true and the false. If he were to attempt to root out the tares, he would in many instances pull up the wheat also. The Lord alone sees the heart, and he knows those who are his. He will, at the great harvest day, separate the one from the other.

## LILY—OLD TESTAMENT.

(*Nymphæa lotus, Linn.*)

---

"My beloved...feedeth among the lilies."—SONG OF SOL. ii. 16, vi. 3.

THE Hebrew word *shushan*, or *shoshannah*, is translated "lily" in the authorized version of the Old Testament. The plant appears to be different from the *krinon*, or lily of the New Testament, although as to this authors are not agreed. Dr. Royle and others consider the lily of the Old Testament to be *Nymphæa lotus*, one of the water-lilies of the Nile. The plant belongs to the natural order Nymphæaceæ. It is the lotus of the ancient Egyptians, sacred to Isis; but it is quite different from the lotus of the Lotophagi, and from the lotus of Homer and Dioscorides, as well as from that of Hippocrates. Its flowers are large, and they are of a white colour, with streaks of pink. They supplied models for the ornaments of the pillars and the molten sea, as described in 1 Kings vii. 19, 22, 26, and 2 Chronicles iv. 5.

The plant grows in still waters and slow-running streams; there it produces its large shield-like leaves, expands its blossoms, and sends forth its fragrant odour. It grows near the margin of the water, two to three feet deep usually. Its large roots are embedded in the mud below. The blade of the leaf is nearly circular, varying

EGYPTIAN LOTUS.

from nine to twenty-four inches in diameter, and attached to the leaf stalk in the very centre. In the young state its rounded leaves float on the surface of the water, but when advanced they rise to the height of four or five feet above the surface. It is a native of Egypt, and is found in the Nile, especially near Rosetta and Damietta, and in rice-fields during the time they are under water.

In the Song of Solomon constant allusion is made to the lilies. Their beauty and their perfume are made to shadow forth the preciousness of Christ to his Church. Thus, in chapter ii., verse 1, Christ says, "I am the rose of Sharon, and the lily of the valleys;" again, in chapter v., verse 13, it is said of Christ by his people, "His lips are like lilies, dropping sweet-smelling myrrh." Hear again what Christ says of his Church: "As the lily among thorns, so is my love among the daughters" (chapter ii., verse 2)—as a glorious and sweet flower in the midst of a thorny wilderness, where all else is bleak and desolate.

This plant is one of those which are alluded to in ancient times as giving a supply of food. The seeds were used to make bread, and the root was also eaten. Even at the present day, in Eastern countries, the roots and stalks furnish articles of diet, and the large farinaceous or mealy seeds of this and another kind of water-lily are roasted and eaten. This may, perhaps, explain the allusions made in the Song of Solomon (ii. 16, iv. 5, and vi. 3), to feeding among the lilies; or the allusion may refer to a kind of cyperus or rush, of which cattle are very fond, and which grows along with the lily in the waters. Christ leads his people here beside still waters, such as those in which the lily grows; and he feeds them with the bread of life.

In Ecclesiastes xi. 1, it is said, "Cast thy bread upon the waters: for thou shalt find it after many days." This may be in allusion to the mode in which the seeds of the lily are sown. They are enveloped in clay and cast into the water; they then sink into the mud, and after many days appear above the water, bearing flowers, and producing seeds, which are used as bread. This mode of sowing is practised now by certain tribes in India.

Hosea says, "Israel shall grow as the lily" (xiv. 5). As the water-lilies grow vigorously in the waters under the shining of the southern sun, so Israel, fed by the refreshing streams of living water, shall flourish under the shining of the Sun of Righteousness.

In the titles of Psalms xlv. and lxix. the word *shoshannim* occurs, which has been translated "lilies." Some have thought that the word refers to the form of the musical instrument used,—resembling lilies; others remark that the imagery in these psalms is considered in part as having reference to what took place at marriages in Egypt, when the female attendants adorned their head-dresses with the water-lilies. How emphatically, then, do these emblems, taken from the lilies, bring out the meaning of the various allusions in the Song of Solomon to Christ as the Bridegroom and his Church as the Bride!

## LILY—NEW TESTAMENT.

(*Anemone coronaria, Linn.*)

---

"Consider the lilies of the field, how they grow."—MATT. vi. 28.

THE word *krina* is translated "lilies" in the New Testament. It occurs in two passages (Matt. vi. 28, and Luke xii. 27), in which our Lord calls upon us to "consider the lilies of the field." There is some difficulty in determining what the plants were. They must have been well known to our Lord's hearers, as growing in the fields near the Sea of Galilee, where he was discoursing. It would appear, from the report of those who have visited Palestine, that in the early spring months the fields abound in various species of lily, tulip, narcissus, iris, and gladiolus; and it is, no doubt, to one of these that reference is made. Many have thought that the white lily (*Lilium candidum*) is the plant referred to; but Royle thinks that this cannot be the case, inasmuch as that plant is not considered to be a native of Palestine; although it is occasionally cultivated there. He is disposed, after careful examination, to conclude that the chalcedonian, or scarlet martagon lily (*Lilium chalcedonicum*), is the " lily of the field." It comes into flower at the season of the year when our Lord's sermon on the mount is supposed to have been delivered; it is abundant in the district of Galilee; and its fine scarlet flowers render it a very conspicuous and showy object, which would naturally attract the attention of his hearers.

LILY OF THE NEW TESTAMENT.

The plant belongs to the natural order Liliaceæ. The six parts of the perianth are of a scarlet colour, and are turned back. Dr. Thomson describes a plant in Palestine, called Hûleh lily, which delights much in the valleys, but is also found on the mountains. It abounds in the woods north of Tabor.

Tristram considers the *Anemone coronaria* as the New Testament *krinon*. He says: "The true floral glories of Palestine are the pheasant's eye (*Adonis*), the ranunculus, and the anemone, but especially the latter. The *Anemone coronaria*, well known in our gardens, of various colours—lilac, white, and red, but most generally a brilliant scarlet—is the flower which is the most gorgeously painted, the most conspicuous in spring, and the most universally spread of all the floral treasures of the Holy Land; and I should feel inclined to fix on it as the 'lily of the field' of our Lord's discourse. It is found everywhere, on all soils, and in all situations. It covers the Mount of Olives, it carpets all the plains, and nowhere does it attain a more luxurious growth than by the shores of the Lake of Galilee." As rivals to the anemone in brilliancy are the *Ranunculus asiaticus* and *Adonis palestina*.

Our Lord refers to the lily as being near the Plain of Gennesaret. It was showy, and it was compared with the robes of Solomon. Probably red, from its comparison to lips. The term "lily" seems to be applied to a number of plants. No doubt there is a lily in the Holy Land (*Lilium chalcedonicum*), the red Turkish lily. Many plants get the name of lily, such as species of iris, gladiolus, narcissus, asphodelus, squills, fritillaria.

We are told that "Solomon in all his glory was not arrayed like one of these" lilies (Matt. vi. 29; Luke xii. 27). In order to understand this, let us look at the beautiful structure in which the colours of the flower reside. The flower-leaves of the lily, when magnified by the microscope, are seen to consist of a number of beautiful honeycomb-like cells, in which the colouring matter is formed and stored. It is of an elegant texture, far exceeding in beauty anything that man could make. Solomon's robes, if examined by means of a magnifying glass, would, so far as they were the work of man, have appeared coarse; but the more the clothing of the lily is magnified, the more exquisite is its beauty. The colours of Solomon's robes might have been gorgeous, but they were not disposed in the way in which God paints the flowers. What are the greatest works of men when compared with those of the Almighty Creator? The green covering of the "grass of the field," which probably means the foliage of the lilies, defies all the art of man to imitate.

How wondrous is the quiet growth of the lilies! There is no toiling or spinning on their part. The process of growth is carried on by an unseen power, even by God, who waters the ground, and who superintends the formation of every minute cell and tube which enter into the composition of the plant.

# MELON.

*(Cucumis melo, Linn.)*

"We remember...the cucumbers and the melons."—Num. xi. 5.

THE Hebrew plural word *abbatichim*, or *abbatichin*, occurs only once in the Bible, and has been translated "melons." It is the *pepones* of the Greek and Latin. The Septuagint has the word *sieyos*, a term which is now-a-days given to a genus allied to the cucumber. The plant referred to is the *Cucumis melo* of botanists, the common melon; and perhaps also *Cucurbita citrullus* (*Cucumis citrullus* of Linnæus), the water-melon. These plants belong to the natural order Cucurbitaceæ, the Cucumber family, which includes sixty-six known genera, and about three hundred and thirty species. The plants of this family are herbs with succulent stems, climbing by means of lateral tendrils which are transformed stipules; their leaves are palmate and rough; their flowers generally unisexual; their stamens five, adhering to the calyx; and their fruit formed by three carpels, constituting what has been called a pepo. The plants are generally acrid in their qualities, although many of them, especially under cultivation, yield edible fruit, such as the cucumber, melon, gourd, pumpkin, squash, and vegetable marrow. Colocynth and elaterium, which are powerful purgatives, also belong to this order. (See article on *Wild Gourd*.) The plants are natives of warm climates chiefly, and abound

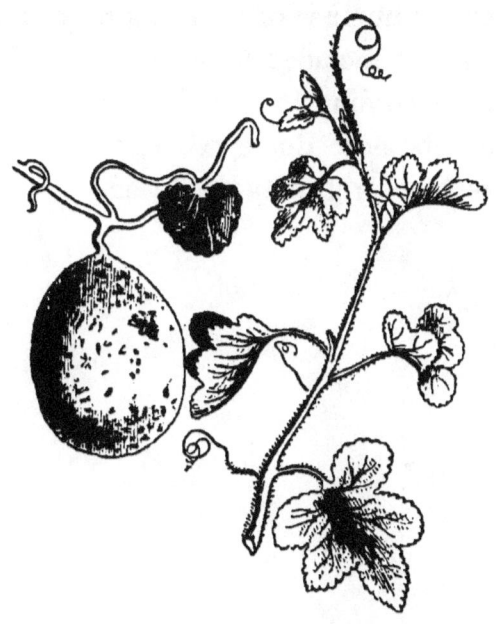

MELON.—(*Cucumis melo.*)

in India. In these countries their edible fruits are highly prized, and hence the words in which the children of Israel alluded to them when they murmured in the desert: "We remember the fish, which we did eat in Egypt freely; the cucumbers, and the melons," etc. (Num. xi. 5.)

Dr. Royle thinks that the common melon is the plant alluded to; and he grounds his opinion in part on the resemblance between the Arabic word for melon, *butikh*, and the Hebrew word. Moreover, he thinks that there is no evidence of the water-melon having been known to the ancient Egyptians. In Arabic the water-melon is called *butikh-hindee*, or Indian melon.

The melon was introduced into Britain about 1520. There are a great number of varieties now in cultivation. The best kinds are included under the name Cantaloupe; an appellation, according to Don, bestowed on them from a seat of the Pope near Rome, where this variety is supposed to have been originally produced.

# NETTLE.

### (*Urtica urens, Linn.*)

---

"It was all grown over with thorns, and nettles had covered the face thereof."—Prov. xxiv. 31.

HE Hebrew words *charul, kimosh,* and *kimshon,* occur in several places in the Old Testament, and have been translated "nettles." There are some doubts as to the correctness of the translation. *Charul* is found in three passages. In Prov. xxiv. 30, 31, it is written: "I went by the field of the slothful, and by the vineyard of the man void of understanding; and, lo, it was all grown over with thorns, and nettles had covered the face thereof." Job says, when speaking of the children of the destitute: "Among the bushes they brayed; under the nettles they were gathered together" (Job xxx. 7). And the prophet Zephaniah, in speaking of the desolation coming on Moab and Ammon, predicts thus: "Surely Moab shall be as Sodom, and the children of Ammon as Gomorrah, even the breeding of nettles, and salt pits, and a perpetual desolation" (Zeph. ii. 9). The plant referred to is

## NETTLE.

obviously one which grows as a weed in gardens, and comes up in desolate places where men have had their habitations. This is very characteristic of the nettle, which follows man's footsteps in all parts of the world—Europe, Asia, Africa, and America. In wild and deserted glens, the sites of cottages are often marked by nettles; and the ruins of old castles give rise to a large crop of these weeds. Some have supposed that a thorny or spiny shrub was meant. Royle is disposed to think, from the resemblance between *charul* and the Arabic *khardul*, that a kind of mustard was referred to, and he has figured *Sinapis orientalis* as the probable species. But this plant does not answer well to the description, seeing it is not a weed of gardens nor a product specially of ruins.

NETTLE.—(*Urtica urens.*)

There is much conjecture on this matter. The correspondence between the word in Hebrew and Arabic no doubt adds plausibility to Royle's conjecture, and our own word charlock is applied also to a kind of mustard. Some of the species of sinapis grow to a great height.

Again: the Hebrew words *kimosh* and *kimshon*, or *kimmashon*, translated "nettle," occur in two places. In

Isaiah xxxiv. 13, it is said, "And thorns shall come up in her palaces, nettles and brambles in the fortresses thereof." Hosea says, "The pleasant places for their silver, nettles shall possess them: thorns shall be in their tabernacles" (ix. 6); and in Proverbs the garden of the sluggard is described as covered with *kimshonim*, translated "thorns," and in some versions "thistles" (xxiv. 31).

Nettles belong to the genus Urtica. There are two common species, which are found generally distributed over the globe;—*Urtica dioica*, the great nettle; and *Urtica urens*, the small nettle. *Urtica pilulifera*, the Roman nettle, also occurs in many places. These belong to the natural order Urticaceæ. They have inconspicuous green flowers without a corolla, their stamens are often elastic, their fruit is a single-seeded nut, and they are covered with stinging hairs.

In Scripture, nettles are made to point out the effect of sloth and idleness, and they indicate the passing nature of all human greatness as regards earthly habitations. They are constantly mentioned as marks of waste and desolation.

# GARLIC.

### (*Allium sativum, Linn.*)

---

"We remember...the leeks, and the onions, and the garlick."—Num. xi. 5.

GARLIC.
(*Allium sativum.*)

THE Hebrew plural word *shumim* occurs only once in the Old Testament, and is translated "garlick." It is the *skordon* of the Greeks. It was one of the vegetable luxuries of Egypt after which the Israelites lusted in the desert. An allied Arabic word is at the present day used for garlic. The plant is the *Allium sativum*, belonging to the natural order Liliaceæ, the Lily family. It was much cultivated in Egypt, and is noticed by Herodotus as having been used in part for the food of the labourers engaged in building the Pyramids. Royle thinks it probable that the shallot (*Allium ascalonicum*) might be the species referred to, and not the common garlic. The shallot, or eschalot, is common in Eastern countries, and derives the name ascalonicum from having been brought into Europe from Ascalon in Palestine.

# GRASS.

SEVERAL Hebrew words are translated "grass." The word *chatzir*, which is commonly applied to the leek, has been also translated "grass." (See article *Leek* and reference to the passage.) The word seems to have been applied to every green herb which could be used for pasture. In many passages of Isaiah the word is used: "All flesh is grass" (Isa. xl. 6); "They shall spring up as among the grass" (Isa. xliv. 4). (See also Isaiah xxxv. 7, li. 12.) There are a great number of grasses in Palestine.

The word *yered* in Numbers xxii. 4 is also translated "grass." *Desher* is the most common name for green grass. It is used in Genesis i. 11: "Let the earth bring forth grass." The name being general includes many species of grass, some being tall and luxuriant, others short and stunted.

Reference is often made in Scripture to the life of man as being like grass: "But the rich, in that he is made low: because as the flower of the grass he shall pass away. For the sun is no sooner risen with a burning heat, but it withereth the grass, and the flower thereof falleth, and the grace of the fashion of it perisheth: so also shall the rich man fade away in his ways" (James i. 10, 11). Here the withering of the grass is used as an emblem of man passing away; and the flower of grass is also alluded to. In the flower of grass, the most conspicuous organs are the stamens, which commonly hang out of the surrounding scales when the plant is in flower. The upper parts of these stamens are attached to slender threads, and the wind easily blows them away. This is very emblematical of the thread of life, which is so easily snapped in a moment.

The flat roofs of the houses at Rûno in Lebanon are constructed, according to Robinson, by laying first large beams at intervals of several feet; then rude joists; on which again are arranged small poles close together, or brushwood; and upon this is spread earth or gravel rolled hard. Grass is often seen growing on these roofs. (Psalm cxxix. 6.)

# LEEK.

(*Allium porrum, Linn.*)

"We remember...the cucumbers, and the melons, and the leeks."
NUM. xi. 5.

THE Hebrew word *chatzir, chazir,* or *chajir,* occurs frequently in the Old Testament, and has been translated "leeks" in Numbers xi. 5, where the Israelites are represented as sighing for the good things of Egypt. The word, however, is rendered differently in other places. Thus in 1 Kings xviii. 5; 2 Kings xix. 26; Job xl. 15; Ps. xxxvii. 2, xc. 5, ciii. 15, civ. 14, cxxix. 6, cxlvii. 8; Isa. xxxvii. 27, xl. 6–8, xliv. 4, li. 12, it is translated "grass;" in Job viii. 12, it is rendered "herb;" in Prov. xxvii. 25, and Isa. xv. 6, it is by mistake translated "hay;" and in Isa. xxxiv. 13, it is rendered "court." The word is derived from a root which means "to be green," and hence it is considered as referring to a green vegetable like grass; and it is probable that the word "court" in Isaiah may have reference to a sort of pasture court. The most ancient Greek translators use the word *prasa* or "leeks" to represent the Hebrew term *chatzir;* and it seems likely that in Numbers, from its association with onions, leeks might be intended, more especially as these vegetables were commonly used at that time in Egypt.

The plant which supplies the leek is the *Allium porrum* of botanists. It belongs to the natural order Liliaceæ, the Lily family. It has grass-like leaves, and its flowers occur in rounded heads. The plant was used as a seasoning to soups in the time of the Romans. It is indigenous in the countries bordering on the Mediterranean. In Egypt it thrives well, and the inhabitants eat their leek and barley bread with avidity. It was introduced into Britain in 1562. The leek was sacred in Egypt, and some have suggested that it was not likely the Israelites would be permitted to eat it there. Lady Callcott says that these plants never were objects of general worship. "They were for the most part reverenced on account

LEEK.—(*Allium porrum.*)

of their being dedicated to, or symbolic of, some well-known deity; much in the way in which a Welshman reverences his leek, the emblem of Wales, and wears it on St. David's Day. That compliment paid, however, he would never think of denying himself the pleasure of eating his leek; and no doubt the ancient Egyptians and their bondsmen made equally free with their savoury gods."

# ONION.

### (*Allium cepa, Linn.*)

---

"We remember...the cucumbers, and the melons, and the leeks, and the onions."—NUM. xi. 5.

THE Hebrew plural word *'betzalim* occurs in Numbers xi. 5, and has been translated "onions." There seems to be no doubt of the correctness of the rendering. The Arabic word is *basl* or *bassal*, which is nearly allied to the Hebrew *betzal*; and it has been rendered by the Greek word *krommyon*, applied to the onion. The plant is the *Allium cepa* belonging to the natural order Liliaceæ, the Lily family. It is a bulbous plant, having its bulbs covered with brown scales; its leaves are tubular or hollow, and its flowers are produced in rounded clusters. It has stimulant, acrid, and pungent qualities, and has been long cultivated in the south of Europe and in the north of Asia.

The Egyptians had a superstitious veneration for onions. When onions become very large, as in Portugal, they lose much of their acrid qualities, and become bland articles of food when cooked. Hasselquist says: "Whoever has tasted onions in Egypt must allow that none can be had better in any part of the universe." While the superstitious inhabitants of Egypt reverenced these productions of the soil, the children of Israel lusted after them in the desert, and murmured against the Lord who had delivered them, and who could supply all their need.

ONIONS.—(*Allium cepa.*)

# WHEAT.

(*Triticum sativum, Linn. ; var. compositum.*)

---

"A land of wheat and barley."—DEUT. viii. 8.

THE Hebrew word *chittah* occurs in many places of the Old Testament, and has been properly translated "wheat." There is also another word, *kemach*, which means "flour of wheat," and which is translated in Genesis xviii. 6 "fine meal." The first distinct notice of wheat in the Bible is in Genesis xxx. 14, where an allusion is made to "wheat harvest." Wheat is a common grain in Egypt, Syria, and other Eastern countries, and is considered as having had an Asiatic origin. It is not known in a wild state. Palestine is spoken of as "a land of wheat" (Deut. viii. 8), and as producing abundantly corn, wine, and oil (Deut. vii. 13, etc.); and the purest wheat or wheat flour is noticed under the name of "the fat of wheat" (Ps. lxxxi. 16, cxlvii. 14, marginal readings), and the kidney-fat, or "the fat of kidneys of wheat" (Deut. xxxii. 14).

Solomon's provision for one day was thirty measures, or cors (probably about a thousand pecks), of fine flour, and threescore measures, or cors, of meal (1 Kings iv. 22). In 1 Kings v. 11, it is mentioned that Solomon gave Hiram twenty thousand measures of wheat for food to his household year by year; and in 2 Chronicles ii. 10, it is stated that a similar amount of beaten wheat was given to Hiram's servants who were employed to cut

WHEAT.—(*Triticum sativum.*)

timber on Mount Lebanon. Wheat from Minnith, a place situated in the domain of the king of Ammon, was famous, and is referred to by Ezekiel as being brought by the Jews to Tyre (Ezek. xxvii. 17). When king Jotham overcame the Ammonites he received from them, as part of the tribute, ten thousand measures of wheat. In the Bible the words corn and parched or roasted corn are frequently used. In many passages they seem to refer to bread corn; that is, wheat.

The common wheat is *Triticum vulgare*, the variety called *Triticum æstivum*, or spring wheat, being sown in spring, and that called *Triticum hybernum*, winter wheat, being sown in autumn. The plant belongs to the natural order Gramineæ, or the Grass family. There are numerous varieties of wheat in cultivation. In Pharaoh's dream the seven ears on one stalk appear to refer to the variety of wheat commonly cultivated in Egypt, and called *Triticum compositum*. This branching variety of wheat helps to explain the allusion in Genesis xli. 5–7, 22–24, and 27. Grains of wheat are found in mummy cases in Egypt, but there is no evidence that any of those *put in along with the mummy*

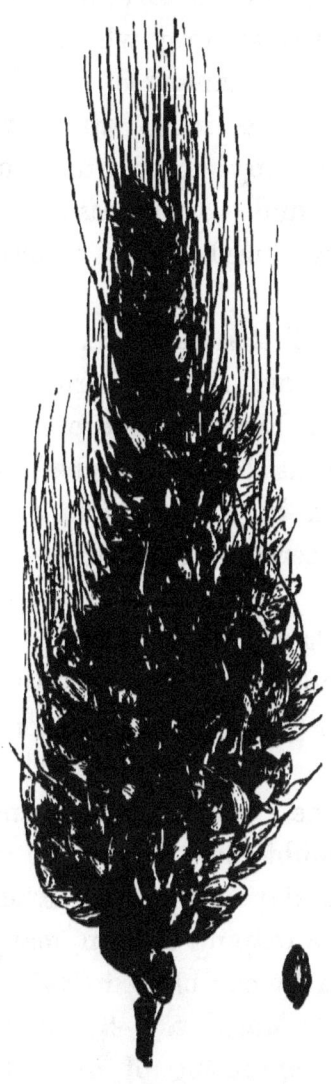

WHEAT.—(*Triticum compositum.*)

have retained their vitality. Grains taken from the cases have no doubt germinated, and in Britain there are many fields of what is called mummy-wheat, but in all instances the grains have been tampered with by guides, who have an interest in deceiving travellers. Among some of the so-called mummy-wheat, grains of Indian corn have actually been found. In no case have any of those mummy-grains produced the *Triticum compositum*.

Reference is made in Leviticus ii. 14 and xxiii. 14 to green ears of corn cut before they are ripe and dried by the fire. It is said that in Lower Egypt, at the present day, such green ears are used as food. Parched or roasted corn is frequently eaten in Eastern countries. When the children of Israel entered Canaan, they ate parched corn (Joshua v. 11). This was one of the articles of food brought to David in the camp at Mahanaim (2 Sam. xvii. 28); and it was given to Ruth by Boaz when she sat beside the reapers at their meal (Ruth ii. 14).

*Corn* is often referred to in the New Testament, and under this name wheat was no doubt included, as well as other kinds of grain, such as barley and spelt. From the sowing, the sprouting, and the reaping of corn, many important illustrations are drawn by our blessed Saviour and his apostles (Matt. xiii. 3–23; Mark iv. 3–20; John xii. 24). St. Paul employs the sprouting of grain to illustrate the believer's resurrection body* (1 Cor. xv. 36–44). Shibboleth, the word which was put as a test

* See Balfour's "Botany and Religion," 3rd edition, p. 51, *et seq.*

to the Ephraimites (Judges xii. 6), is the Hebrew name for an ear of corn.

Parched corn, under the name of *kali*, is referred to in several passages of Scripture, as Lev. xxiii. 14; Ruth ii. 14; 1 Sam. xvii. 17, xxv. 18. Some have supposed the kali referred to the produce of the chick-pea, *Cicer arietinum*. Dr. Thomson says that at the present day parched corn is used during harvest. "It is made thus: A quantity of the best ears, not too ripe, are plucked with the stalks attached; these are tied into small parcels; a blazing fire is kindled with dry grass and thorn bushes, and the corn heads are held in it until the chaff is mostly burned off. The grain is thus sufficiently roasted to be eaten, and it is a favourite article all over the country. ...After it has been roasted, it is rubbed out in the hand and eaten as there is occasion." (*The Land and the Book*, p. 648.) The green ears of corn are also constantly plucked and rubbed in the hands (Matt. xii. 1, 2; Mark ii. 23; Luke vi. 1, 2); and the taking of them is not considered an act of stealing.

# SPELT.

### TRANSLATED "RYE."

(*Triticum Spelta, Linn.*)

---

"The appointed barley and the rie [spelt] in their place."—ISA. xxviii. 25.

 A KIND of wheat called *spelt* (*Triticum Spelta* of botanists) seems to be referred to under the Hebrew name of *kussemeth*, which has been translated "rye" in Exodus ix. 32, Isaiah xxviii. 25, and "fitches" in Ezekiel iv. 9. Rye is a grain of cold climates, and is not cultivated in the southern parts of Europe.

*Kussemeth* was undoubtedly one of the cultivated crops of Egypt and Syria, and was used as an article of food. It seems to have been sown at the same time as wheat, and is referred to in the seventh plague of Egypt as not having been smitten, because, like the wheat, it was not grown up. Ezekiel mentions it as being used in making bread. Some have supposed that the spelt was sown as a border round other kinds of grain, and that allusion is made to this in Isaiah xxviii. 25 (marginal reading). Spelt is a bearded kind of wheat, and in this respect has a resemblance to rye. The names *olyra* and *zea* were given to it by some Greek authors. It is cultivated in the south of Germany. The plant belongs to the natural order Gramineæ, the Grass family.

# BARLEY.

*(Hordeum distichon, Linn.)*

---

"A land of wheat and barley."—DEUT. viii. 8.

BARLEY is another kind of grain mentioned both in the Old and in the New Testament. It is referred to under the Hebrew name of *seorah* or *shoreh*, and under the Greek name of *krithe*. It is the *Hordeum distichon* of botanists, and belongs to the natural order Gramineæ, the Grass family. It is mentioned along with common and spelt wheat. Oats and rye, being northern plants, did not grow in Palestine. The two-rowed barley is that which is most commonly cultivated. *Hordeum vulgare*, bere, big or four-rowed barley, and *Hordeum hexastichon*, six-rowed barley, are confined to higher regions, and are not commonly cultivated in Britain. The bere, however, finds a place in the present fiars of upwards of twenty counties in Scotland. Barley is one of the most ancient articles of diet. It is often noticed along with wheat as occurring in Palestine, and as having been used for food (Deut. viii.

8; 2 Chron. ii. 10, 15, xxvii. 5). Barley was grown by the Egyptians and the Jews, and was used for making bread and cakes. It was mixed also with wheat, lentiles, and millet. In 1 Kings iv. 28, barley is mentioned as having been used as food for Solomon's horses. Barley meal was employed in certain instances as an offering (Num. v. 15). Barley bread served as food for the common people; and the loaves which were miraculously distributed to the multitude by our Lord were made of barley (John vi. 9, 13). The friends of David brought barley to him when he fled from Absalom (2 Sam. xvii. 28). Barley harvest is mentioned in Ruth i. 22, ii. 23; and 2 Samuel xxi. 9, 10. This takes place in Palestine about the end of March or the beginning of April. The barley ripens in Egypt about a month before the wheat; and hence it was destroyed by the hailstones, while the wheat escaped (Ex. ix. 31). Boaz measured six measures of barley, and put it into Ruth's veil (Ruth iii. 15). This veil was consequently made of stronger material than veils in this country.

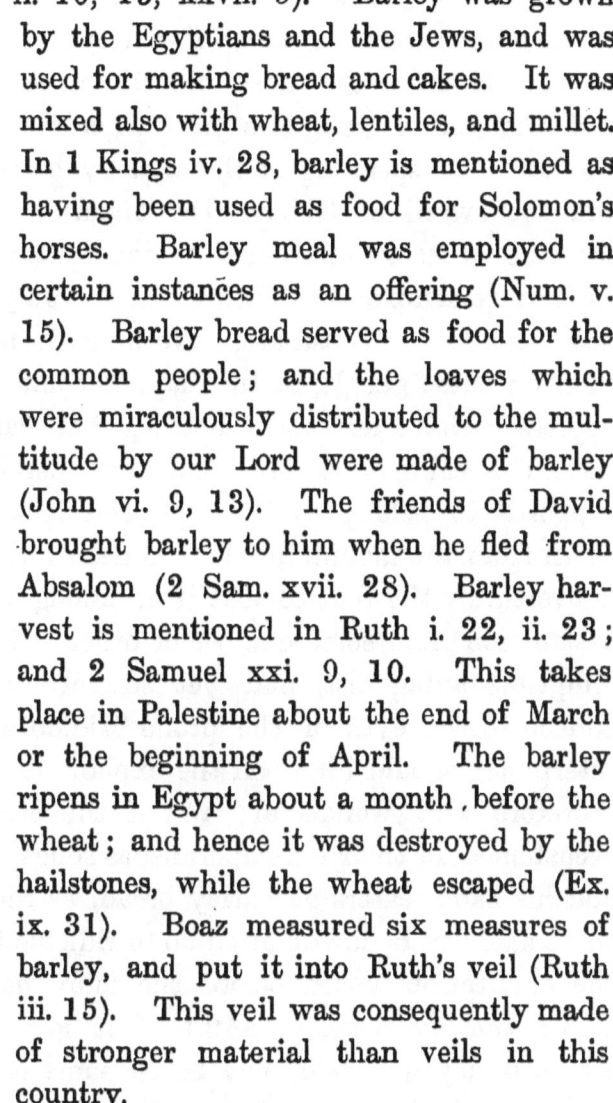

BARLEY.
(*Hordeum distichon.*)

Barley bread was not much esteemed by the Jews. Ezekiel says (xiii. 19), "Will ye pollute me among my people for handfuls of barley?"—probably referring to its small value. In Gideon's dream a cake of barley bread is observed to tumble into the host of Midian and smite it; and the man's fellow says, "This is the sword of Gideon." In speaking of the analogy between the cake and the sword of Gideon, Dr. Thomson says: "As to the line of connection in the mind of the 'interpreter,' we may remember that barley bread is only eaten by the poor and the unfortunate. Nothing is more common than for these people, at this day, to complain that their oppressors have left them nothing but barley bread to eat...This cake of barley bread was therefore naturally supposed to belong to the oppressed Israelites: it came down from the mountain where Gideon was known to be; it overthrew the tent so that it lay along, foreshadowing destruction from some quarter or other. It was a contemptible antagonist, and yet scarcely more so than Gideon in the eyes of the proud Midianites. That the interpreter should hit upon the explanation given, is not therefore very wonderful; and if the Midianites were accustomed, in their extemporaneous songs, to call Gideon and his band 'eaters of barley bread,' as their successors, these haughty Bedawîn, often do to ridicule their enemies, the application would be all the more natural." (*The Land and the Book*, p. 449.) The low estimation in which barley was held may be in some way implied in its use in the jealousy-offering (Num. v. 15).

# COCKLE.

*(Hebrew, Baoshah.)*

---

"Let thistles grow instead of wheat, and cockle instead of barley."
JOB xxxi. 40.

HE Hebrew word *baoshah*, translated "cockle" in the Authorized Version, is rendered "noisome weeds" by some. This is the marginal reading in Job. In Isaiah v. 2, 4, *benshin*, the plural of the word, occurs, and is translated "wild grapes."

Celsius traced the word to *biseh*, which is applied to a kind of monkshood. Its name seems to have been applied to a troublesome weed with an offensive smell. The author of the "Scripture Garden Walk" says: "The Greek word adopted in the Septuagint implies a bush, or bramble, bearing berries. According to Castalio, the tree was the dwarf elder; and other commentators have supposed other senses of the word." Tristram suggests some species of arum which grows in the fields of Palestine. There is no satisfactory translation of the word.

# GOURD.

(*Ricinus communis, Linn.; Cucurbita pepo, Linn.*)

---

"The Lord prepared a gourd, and made it come up over Jonah."
JONAH iv. 6.

THE Hebrew word *kikayon*, translated "gourd," occurs in the fourth chapter of the Book of Jonah, verses 6, 7, 9, and 10. It is probably the *kiki* of the Greeks, which is described as a plant having leaves like a palm-tree, and seeds which yield oil, contained in a rough seed-vessel. In character and properties the plant corresponds with the castor-oil plant, the *Ricinus communis* of botanists. The marginal reading gives "palm-crist," which seems to be a corruption of *Palma Christi*, another name for the castor-oil plant. It belongs to the natural order Euphorbiaceæ, the Spurge family. It is a native of Southern Europe, Palestine, and India. In Europe generally it is cultivated in greenhouses as an herb, and does not attain a large size; but in India and other warm climates it becomes arborescent, so as to afford a shelter from the sun's rays. The stem of the plant is usually soft, and is easily destroyed by insects or worms. In China a peculiar fungus called *Hirneola auricula-Judæ*, used in soups, grows on the decaying stems of the ricinus. The seeds of the plant when bruised yield the oil called castor oil. The name of *kik oil* is also applied to it. It is said that the modern Jews in London use this oil for their Sabbath lamps.

CASTOR-OIL PLANT.—(*Ricinus communis.*)

In place of the castor-oil plant it may be that the common gourd (*Cucurbita pepo*) is meant. Tristram says that in Palestine the name for the gourd is *kurah*, while that for the castor-oil tree is *khurwah*, so that the sound of each is nearly identical, and hence there might be a confusion as to the name on the part of the commentators.

The gourd is constantly grown in Palestine now to cover arbours, and its rapid growth and large leaves render it very well fitted for affording shade. Jonah is said to have made him a booth, and God prepared the gourd to cover it. This undoubtedly seems rather to confirm the idea that the common gourd and not the castor-oil plant is referred to.

Many have been the disputes as to Jonah's gourd, and it is impossible to decide the point with certainty. The castor-oil plant seems, upon the whole, to fulfil the conditions required; and the cognate word in Greek helps to decide the matter in some measure. The gourd was prepared by the Lord miraculously for Jonah, in order that its broad leaves might be a shadow over his head, to protect him from the heat; and the prophet, we are told, was "exceeding glad of the gourd" (Jonah iv. 6). It was a temporal blessing provided for him; but, alas! like all creature comforts, it was fleeting. By the time the sun rose next morning, a worm, by God's command, smote the gourd that it withered. The destruction of the gourd, and the removal of the shade, made Jonah angry, and he wished in himself to die, and said, "It is better for me to die than to live" (Jonah iv. 8). He grumbled at the orderings of God's providence, and his proud heart rose against God's dispensations. The lesson which God conveyed to Jonah is thus expressed: "Then said the Lord, Thou hast had pity on the gourd, for the which thou hast not laboured, neither madest it grow; which came up in a night, and perished in a night: and should not I spare Nineveh, that great city, wherein are more than sixscore thousand persons that cannot discern between their right hand and their left hand; and also much cattle?" (Jonah iv. 10, 11.)

# CUCUMBER.

(*Cucumis sativus, Linn.*)

---

"We remember...the cucumbers and the melons."—NUM. xi. 5.

THE Hebrew word *kishuim* occurs twice in the Old Testament, and has been translated "cucumbers." The singular of the word is *kisha*, which resembles the Arabic *kissa*, the name for cucumber. In Greek the name is *sicyos*. It is the *Cucumis sativus*, and belongs to the natural order Cucurbitaceæ, the Cucumber and Gourd family. It is called *ketimou* and *timou* by the Hindus. It is a native of Eastern countries, and was introduced into Britain in 1573. It is a trailing and climbing plant, with large rough leaves, having tendrils. The plant was known in very early times, and it was cultivated extensively in Egypt. Hence the allusion made by the children of Israel as recorded in Numbers xi. 5. They longed for the cucumbers of Egypt. There are a great number of varieties in cultivation. A species called *Cucumis chate* grows near Cairo after the inundation of the Nile, and is said to yield a delicious fruit highly esteemed in

Egypt. This was probably used along with the common cucumber, and is included in the Hebrew word. Egypt may still be called a land of cucumbers. The guarding of vineyards and cucumber beds is referred to by the prophet Isaiah when he describes the desolation of Israel: "The daughter of Zion is left as a cottage in a vineyard, as a lodge in a garden of cucumbers" (Isa. i. 8). Lady

CUCUMBER.—(*Cucumis sativus.*)

Callcott remarks: "This statement of the prophet is constantly recalled to the memory of the modern traveller in Egypt by the vast plantations of cucumbers on the banks of the Nile. There, as of old, the peasant has his lodge, that he may water his rich plants with the *shadoof*, or, as the scripture expresses it, 'by the foot' (Deut. xi.

10); and that he may guard his little property from the robbers of the Nile, who, though of a different class, are not less formidable to the cultivators than those of the time of Herodotus." They all require to be protected from wild animals, such as the jackals. The lodge is stated to be constructed rudely of four poles, with rafters laid across these; and branches cut from oleanders, or matting, put on the top, to give shelter. When the cucumbers are gathered, and the field is deserted, Dr. Thomson says the poles fall down and lean every way, and the green branches are scattered by the wind, presenting a type of utter desolation. This covering is the " booth which the keeper maketh," mentioned in Job xxvii. 18. The young green fruit of the cucumber is preserved as a pickle, under the name of *gherkins*, which is a corruption of the German word *gurke*, meaning cucumber.

# BULRUSH AND RUSH.

(*Papyrus antiquorum,* Willd.)

---

"Can the rush [bulrush] grow up without mire?"—Job viii. 11.

THE Hebrew words *gome* and *agmon* occur in several passages in the Old Testament, and have been translated "bulrush," and "rush," and "flag." The word *gome* means originally to soak or drink up, and it is therefore given to a plant growing in watery and marshy places. In Isaiah xxxv. 7, it is noticed as a plant of wet places. It is supposed to be the *Papyrus antiquorum,* which grew in large quantity in Egypt among the mud of the Nile. The plant has entangled, spreading roots, and underground stems, which cause the mud to accumulate; and by forming a more or less solid clay it seems to drink up the water in which it grows. The plant appears to have contributed in no small degree to form the Delta of the Nile, and in so doing it has become extinct, from want of wet mud in which to grow. In Job viii. 11, it is said, "Can the rush grow up without mire?" It is found still in the marshes of the White Nile in Nubia.

Tristram says that it is found in two places in Palestine,—in a swamp at the north end of the Plain of Gennesaret, and in the marshes of the Huleh, the ancient Merom. The Arabs call it *babeer*.

PAPYRUS PLANT.—(*Papyrus antiquorum.*)

The plant belongs to the natural order Cyperaceæ, or the Sedge family.

The papyrus was used in Egypt for forming light

sorts of boats; and hence, in Isaiah xviii. 2, "vessels of bulrushes upon the waters" are mentioned. (See also Pliny, xiii. 11, and Lucian, iv. 136.) Jochebed, the mother of Moses, constructed an ark, or little covered boat, of bulrushes, and in this the babe floated on the water of

BULRUSH.

the river (Ex. ii. 3). Boats are at the present day constructed of various kinds of allied plants. Balsas, or fishing-rafts used on the Pacific coast of South America, are formed of the stalks of *Scirpus lacustris*. The name "paper" is derived from the papyrus, which was anciently used in its manufacture. The paper was made by splitting up the stalks of the plant into thin slices of cellular tissue, and then cementing them together. The structure of the paper is the same as rice paper. In Sicily, at the present day, there is a coarse paper made from the papyrus.

The word *agmon* occurs in Isaiah ix. 14 and xix. 15, where it is translated "rush;" in Isaiah lviii. 5, where it is called "bulrush;" and in Job xli. 2, under the name "hook." This last, according to some commentators, should be translated, "Canst thou tie up his mouth with a *rush-rope?*" It seems to have been a kind of reed, but it is not easy to pronounce upon the species. Some consider it is similar to *kaneh*, translated "reed," and look upon it as a species of *arundo* like the common reed, or the variety of *Arundo donax* called *œgyptiaca*.

Royle says that "various species of reeds (*arundo*) will suit the different passages in which this word *agmon* occurs; but several species of *saccharum*, growing to a great size in moist situations, and reed-like in appearance, will also fulfil all the conditions required, as affording shelter for the behemoth or hippopotamus, being convertible into ropes, forming a contrast with their hollow stems to the solidity and strength of the branches of trees, and when dry easily set on fire, and when in flower their light and feathery inflorescence may be bent down by the slightest wind that blows." (See *Reed*.)

# SPIKENARD.

(*Nardostachys jatamansi, Dec.*)

---

"My spikenard sendeth forth the smell thereof."—SONG OF SOL. i. 12.

HE Hebrew word *nerd* or *nard*, and the Greek *nardos*, have been translated in our version of the Bible "spikenard." From the references made in Scripture it is clear that the plant was one famous for its perfume. In the Song of Solomon i. 12, it is said, "While the king sitteth at his table, my spikenard sendeth forth the smell thereof;" and in ch. iv. 13, 14, the plant is mentioned as cultivated in gardens along with "trees of frankincense; myrrh and aloes, with all the chief spices." Many of these are known to have been the products of Arabia and far Eastern countries, and to have been brought to Palestine, especially in the days of Solomon. In Mark xiv. 3, spikenard is referred to both as regards its perfume and its value. While Jesus sat at meat, "there came a woman having an alabaster box of ointment of spikenard very precious; and she brake the box, and poured it on his head." In John xii. 3 the same occurrence is alluded

to: "Then took Mary a pound of ointment of spikenard, very costly, and anointed the feet of Jesus, and wiped his feet with her hair: and the house was filled with the odour of the ointment." The value of the ointment is referred to by Judas Iscariot, who said that it might have been "sold for three hundred pence [denarii], and given to the poor" (John xii. 5). Rosenmüller says this sum was about fifty rixdollars; that is, about eight pounds sterling. All these passages point to the delightful perfume and the rarity and costliness of spikenard. The term *nard* is said to be derived from the Tamil, in which words beginning with *nar* convey the notion of an agreeable perfume. The ointment prepared from the oil of spikenard-root was considered by the Romans as precious. Horace promises to Virgil a whole cadus (about thirty-six quart bottles) of wine for a small onyx-box full of spikenard:—

"Nardo vina merebere,
Nardi parvus onyx eliciet cadum."
Hor., *Carm.* iv., *Ode* 12.

On the occasion of banquets the Romans crowned their guests with flowers such as roses, and anointed them with spikenard:—

"Et rosa,
Canos odorati capillos,
Dum licet, Assyriaque nardo
Potamus uncti."
Hor., *Carm.* ii., *Ode* 11.

Sir William Jones, in the "Asiatic Researches," states that he considers the spikenard like the produce of a plant called in Bengal *jatamansi*, the stem of which, covered with fibrous matter, is dug up in the young state, dried, and sold in the bazaars. In this state it resembles the tail of an ermine or small weasel. The plant has also, from its form, been called by the Arabs *Sunbul hindae,* or Indian ear.

The plant has been specially examined by Dr. Royle. It is the *Nardostachys jatamansi* of Decandolle. It belongs to the natural order Valerianaceæ, the Valerian family. The general name is derived from Greek words meaning "nard" and "spike," and the specific name is from the Indian appellation. This Indian plant seems to have been imported from the Himalayas in the days of Solomon, and to have been prized as a rare kind of perfume. Our Lord smelled a sweet savour when Mary anointed him; and he commended her dutiful faith when he said, in reply to the covetous and hypocritical Judas, "Let her alone; why trouble ye her? she hath wrought a good work on me...She hath done what she could: she is come aforehand to anoint my body to the burying. Verily I say unto you, Wheresoever this gospel shall be preached throughout the whole world, this also that she hath done shall be spoken of for a memorial of her" (Mark xiv. 6, 8, 9).

# COTTON.

*(Gossypium herbaceum, Linn.)*

---

"Where were white, green [cotton], and blue, hangings."—ESTHER i. 6.

THE word cotton does not occur in our translation of the Bible; but there is a Hebrew word, *karpas*, in Esther i. 6, which has been translated "green," and which, according to many commentators, means cotton. The passage in Esther describing the hangings of the palace of Ahasuerus called Shushan or Lily, should, according to these commentators, be rendered thus: " Hangings of white cotton and blue, fastened with cords of fine flax and purple to rings of silver and pillars of marble." The scene of Esther's history was a country where cotton has been constantly used to supply articles of clothing; and this tends to strengthen the opinion that cotton is referred to in the passage which has been quoted. The Hebrew word *karpas* is very like the Sanskrit *karpasum* and *karpasa*, signifying the cotton plant; and it resembles the Latin *carbasus*, which also means cotton. Royle remarks that the hangings thus described in Esther are exactly like those used in India; for "hanging curtains made with calico, usually in stripes of different colours, are employed throughout India as a substitute for doors." The Indian name of cotton might easily reach the Persian court of Susa in the time of Ahasuerus, whose

COTTON.—(*Gossypium herbaceum.*)

dominion extended to India. As the communication between India and Egypt was great, it is probable that cotton was introduced into the latter country. The Jews, in all probability, brought cotton with them on their return from Babylon.

Some authors have supposed that the Hebrew words

*shesh, bad, butz*, and the Greek *byssus*, which occur in the Bible, and which are translated "linen" and "fine linen," may refer to cotton. These views, however, do not appear to have been confirmed by the best commentators.

The common herbaceous cotton is the *Gossypium herbaceum* of botanists. It belongs to the natural order Malvaceæ, the Mallow family. The plant has five lobed leaves, and yellow petals, with a purple spot on each claw. It is found in India, and it also occurs in the south of Europe. There are other species of cotton cultivated in various parts of the world, especially in America.

The substance called cotton consists of the hairs which surround the seeds in the capsule or seed-vessel, and which are the means of scattering the seed when the capsule opens. While God thus wisely provides for the dispersion of the cotton seed, he has also graciously prepared for man materials of a most valuable kind for his clothing and comfort. The history of cotton, its preparation and manufacture, is a subject of deep interest, and is connected with the commercial history of nations. Royle remarks that "cotton has from the earliest ages been characteristic of India. Indeed it has been well remarked, that as from early times sheep wool has been principally employed for clothing in Palestine and Syria, in Asia Minor, Greece, Italy, and Spain; hemp in the northern countries of Europe, and flax in Egypt; so cotton has always been employed for the same purpose in India and silk in China. In the present day, cotton, by the aid of machinery, has been manufactured in this country on so extensive a scale, and sold at so cheap a rate, as to have driven the manufacture of India almost entirely out of the market. But still, until a very recent period, the calicoes and chintzes of India formed very extensive articles of commerce from that country to Europe." Cotton is now largely cultivated in Palestine.

# REED.

(*Arundo donax, Linn.*)

---

"The reeds and flags shall wither."—Isa. xix. 6.

THE Hebrew word *kaneh* and the Greek *kalamos* have been translated "reed" in the Bible. The word *canna* in Greek and Latin, and *cane* in English, may probably be traced to the Hebrew word *kaneh*. In the Old Testament the word *kaneh* is generally applied to reeds growing in water, the hollow stems of which are easily broken. Thus in Isaiah xix. 6, it is said, "And they shall turn the rivers far away...and the reeds and flags shall wither." In Isaiah xxxv. 7, reeds and rushes are associated as growing in water; in 1 Kings xiv. 15, we have the expression, "As a reed is shaken in the water;" Job xl. 21, "He lieth under the shady trees, in the covert of the reed, and fens." The bruised reed is referred to in Isaiah xlii. 3, and other places; and its fragile character is noticed in 2 Kings xviii. 21: "Now...thou trustest upon the staff of this bruised reed, even upon Egypt, on which if a man lean, it will go into his hand, and pierce it: so

is Pharaoh king of Egypt unto all that trust in him."
(See also Isa. xxxvi. 6; Ezek. xxix. 6, 7.) Judging from
the names given by Greek and Roman authors to the
plants of Syria and Egypt, we may conjecture that in
these passages a species of *arundo* was referred to, such
as *Arundo phragmites*, the common reed, or *Arundo
donax*. This plant belongs to the natural order Gramineæ, the Grass family.

In the New Testament the word *kalamos* is used as a
translation of *kaneh* in Matthew xii. 20. This word is
the *calamus* of the Latin; and from it we derive the
term culm, applied to the stems of grasses. In Matthew
xi. 7, and Luke vii. 24, our Lord, in speaking of John
the Baptist, says to the people, "What went ye out into
the wilderness to see? A reed [*kalamos*] shaken with
the wind?" Again, when the Roman soldiers mocked
Jesus, it is said they "put a reed in his right hand," and
they "took the reed, and smote him on the head" (Matt.
xxvii. 29, 30; Mark xv. 19). A reed was used to raise
up the vinegar on the sponge to the lips of the Saviour
on the cross (Matt. xxvii. 48; Mark xv. 36); and John
states that hyssop was also used—probably meaning that
the sponge was put on a bunch of the hyssop shrub
attached to a reed. The apostle John used the word
*calamus* to mean a pen made of a reed (3 John 13).

(See also *Bulrush* and *Flag*.)

# FLAG.

*(Cyperus esculentus, Linn.)*

---

"Can the flag grow without water?"—JOB viii. 11.

THE Hebrew word *achu* occurs in Job viii. 11, and is translated "flag"—"Can the rush grow up without mire? can the flag [*achu*] grow without water?" In this passage the word seems obviously to apply to an aquatic plant of some sort. In Genesis xli. 2, 18, however, the same word has been translated "meadow"—"And, behold, there came up out of the river seven well-favoured kine and fat-fleshed; and they fed in a meadow." In the latter passage the word, according to our translation, embraces the pasture or the moist meadow on the bank of the river on which the flag grew. Commentators think that the flag was a plant of the sedge family, and probably a species of *cyperus* which furnished pasture for cattle. Hence the *Cyperus esculentus*—so called from its esculent qualities—has been conjectured to be the flag of the Bible. It belongs to the natural order Cyperaceæ. The plant grows in the south of Europe, in Africa, and in the

East. It produces tuberous roots which are nutritious, and which, when roasted, have been used as a substitute for coffee. The plant called flowering rush (*Butomus umbellatus*), belonging to a distinct order, is also found in Palestine.

Another Hebrew word, *suph*, has been translated "flag." It is met with in Exodus ii. 5, 6, where Jochebed is represented as placing the little ark with Moses in the flags by the river's brink. Again, in Isaiah xix. 6, it is said, "The reeds and flags [*suph*] shall wither." In Jonah ii. 5 the word is rendered "weeds" —"The depth closed me round about, the weeds were wrapped about my head." Some have supposed that the word is applied to sea-weed in general. Lady Callcott is disposed to look upon it as referring to a species of sea-wrack, such as *Zostera marina*, or *Caulinia oceanica*. The latter plants are thrown up by the tide in the form of numerous balls on the shores of the Mediterranean Sea. The rush-like covering of Florence flasks is made from *zostera*. This plant is not a true sea-weed, but is in reality a flowering plant belonging to the natural order Naiadaceæ. The Red Sea is called Yam Suph, or "the sea of weeds."

# DOVE'S DUNG.

(*Ornithogalum umbellatum, Linn.*)

---

"And the fourth part of a cab of dove's dung was sold for five pieces of silver."—2 KINGS vi. 25.

THE Hebrew *chirionim*, or *charei-yonim*, is met with in the passage in 2 Kings referred to in the above quotation, in which the famine in Samaria was so great that comparatively worthless articles of food were sold for a high price. The word has been translated literally "dove's dung," the term *yonim* being a plural word meaning "doves," and the prefix *charei*, sometimes put *dib*, means "dung." Some commentators believe that the actual dung of pigeons is meant, and that the people were reduced to such straits as to be compelled to eat such offensive materials. It has been stated that in the famine in England in 1316 the poor actually ate pigeon's dung. Other commentators think that the cab of dove's dung is part of a plant which received that name. The Arabs applied the term to certain vegetable productions. A species of *salsola* is called by them "sparrow's dung."

The plant figured has been called the dove's dung plant on account of the green and white colour of its flowers, thus resembling pigeon's dung.

STAR OF BETHLEHEM.
(*Ornithogalum umbellatum.*)

The plant (*Ornithogalum umbellatum*) is said to be abundant in Samaria. It belongs to the natural order Liliaceæ, the Lily family. It grows abundantly in Europe as well as in the Levant. The cab is a measure equal to three English pints. The names of bird's milk and common star of Bethlehem are also given to the plant. As this ornithogalum is found in England, it might supply the pigeon's dung mentioned in the English famine. The bulb is used as an esculent in Syria and neighbouring countries. It was formerly eaten by the peasants in Italy.

# MANDRAKE.

(*Atropa mandragora, Linn.; Mandragora officinalis, Mill.*)

"Reuben...found mandrakes in the field."—GEN. xxx. 14.

THE Hebrew plural word *dudaim* occurs in two passages in the Old Testament, and has been translated "mandrakes." The plant is described as growing in the fields, and as producing its fruit at the time of wheat-harvest, or in May. "Reuben went in the days of wheat harvest, and found mandrakes in the field, and brought them unto his mother Leah. Then Rachel said to Leah, Give me, I pray thee, of thy son's mandrakes. And she said unto her, Is it a small matter that thou hast taken my husband? and wouldest thou take away my son's mandrakes also? And Rachel said, Therefore he shall lie with thee to night for thy son's mandrakes. And Jacob came out of the field in the evening, and Leah went out to meet him, and said, Thou must come in unto me; for surely I have hired thee with my son's mandrakes. And he lay with her that night. And God hearkened unto Leah, and she conceived, and bare Jacob the fifth son." From this passage it appears that the plant was considered as promoting conception. Again, in the Song of Solomon, allusion is made to the smell of the mandrakes: "The mandrakes give a smell, and at our gates are all manner of pleasant fruits" (vii. 13).

There have been numerous opinions as to the plant referred to in these passages. The Greek translators used the word *mandragorai*, "mandrakes," and *Mala mandragoroon*, or "apples of mandrakes," to express the Hebrew *dudaim*. Hence the plant has been considered as the *Atropa mandragora* of botanists. The plant possesses stimulant and narcotic qualities. It belongs to the natural order Solanaceæ, sub-order Atropeæ, or Deadly-nightshade family. The leaves of the plant are coarse and lettuce-like, and they conceal the pale yellowish flowers which arise from the crown of the root. The root is large and spindle-shaped, and often divides in a forking manner. It has a resemblance to the human form, and hence the plant was sometimes called anthropomorphon. The fruit resembles the potato-apple, and is of a pale orange colour. It seems to have been called sometimes "apple of love." The plant occurs in Palestine, and has been noticed by recent travellers. The inhabitants reckon the fruit exhilarating, and as aiding in the procreation of children. There is a cucurbitaceous

MANDRAKE.—(*Atropa mandragora*)

plant to which the name dudaim is given at the present day. This is the *Cucumus dudaim*, or apple-shaped melon, which has a fruit variegated with green and orange at first, and becoming yellow when ripe. The fruit has a very fragrant, vinous, musky odour, and contains a whitish, insipid pulp. Its qualities are very different from those of the mandrake.

There are some curious legends about the mandrake. It was thought that the man-like root, when torn from the ground, uttered shrieks,—

> "Shrieks like mandrakes' torn out of the earth,
> That living mortals, hearing them, run mad."
> *Romeo and Juliet.*

In the Second Part of Henry VI. Shakespeare also alludes to this notion, when he makes Suffolk say,—

> "Would curses kill as doth the mandrake's groan."

In the old time of sorcery and magic the plant acquired a remarkable reputation, and was regarded with superstitious fear.

Its narcotic qualities are referred to by Shakespeare,—

> "Not poppy, nor mandragora,
> Nor all the drowsy syrups of the world,
> Shall ever medicine thee to that sweet sleep
> Which thou owedst yesterday."

Lady Callcott in her "Scripture Herbal" refers to these passages from Shakespeare, and gives some curious details in regard to the plant.

Another solanaceous plant, *Physalis alkekengi*, winter cherry or Jew's cherry, has by some been regarded as yielding the fruit called dudaim.

# THISTLE.

(*Tribulus terrestris, Linn.*)

---

"Do men gather...figs of thistles?"—MATT. vii. 16.

THE Hebrew word *dardar*, and the Greek *tribolos*, have been translated in the Authorized Version "thistles." When God cursed the earth for man's sin, he said, "Thorns also and thistles shall it bring forth to thee" (Gen. iii. 18); and in announcing judgment on Israel, the prophet says, "The thorn and the thistle shall come up on their altars" (Hosea x. 8). In these passages the word *dardar* is associated with *koz* or *kotz*, meaning "thorns." Again, in the New Testament our Lord says, "Do men gather...figs of thistles?" (Matt. vii. 16.) The word *tribolos* or *tribulus* is translated "briers" in Heb. vi. 8: "That which beareth thorns and briers is rejected." There is some difficulty in ascertaining what plant is meant. Some suppose that it is *Tribulus terrestris*, a plant which derives its name from the Greek name *tribolos*. It is a prickly plant which grows along the surface of the ground. It is called *caltrops*, in consequence of the spiny fruit resembling the machines formerly used to obstruct cavalry. It grows in dry barren places in the East. This plant belongs to the natural order Zygophyllaceæ, the Bean-caper family. Some commentators consider the plant as *Centaurea calcitrapa*, one of the composite plants; others take *Fagonia cretica* or *F. arabica*. Lady Callcott figures in her "Scripture Herbal" *Carduus arabicus*, a true species of thistle. There is no doubt that thistles are common in the Holy Land at the present day. Hasselquist noticed eight or ten different kinds of thistles on the road from Jerusalem to Rama, and one on Mount Tabor, along with the *Cynara scolymus*, or the artichoke, which belongs to the same order as the thistle. Other composite plants remarkably spiny, and which are mentioned as occurring in Palestine, are *Cnicus syriacus*, *Scolymus maculatus*, *Centaurea calcitrapa* and *C. verutum*, and *Carthamus oxycanthus*. In bringing forth thistles, the land produces what is highly injurious to cultivation, for the down or pappus attached to the fruit scatters the seeds far and wide, and the plants thus produced choke all useful vegetation. The down of the thistle and of our composite plants is an altered and degenerate calyx. In 2 Kings xiv. 9, as well as in 2 Chronicles xxv. 18, and in Job xxxi. 40, the Hebrew word used for thistles is not *dardar*, but *choach;* and this latter word is also translated "thorn" in Job xli. 2, Proverbs xxvi. 9, Song of Solomon ii. 2, Isaiah xxxiv. 13, and Hosea ix. 6. There can be no doubt that these names refer to noxious weeds which are connected with desolation and a curse.

# HEMLOCK.

(AN UNKNOWN PLANT.)

"As hemlock in the furrows of the field."—HOSEA. x. 4.

THE Hebrew word *rosh* has been variously translated in the authorized version of the Old Testament. In Hosea x. 4, it is translated "hemlock:" "Thus judgment springeth up as hemlock in the furrows of the field." So also in Amos vi. 12: "Ye have turned...the fruit of righteousness into hemlock." In other passages the word is rendered "gall," and it is often associated with wormwood. Thus, in Deuteronomy xxix. 18, it is said, "Lest there should be among you a root that beareth gall and wormwood." The prophet Jeremiah says, "The Lord our God...hath given us water of gall to drink" (viii. 14); "Behold, I will feed them, even this people, with wormwood, and give them water of gall to drink" (ix. 15); "I will feed them with wormwood, and make them drink the water of gall" (xxiii. 15); "Remembering mine affliction and my misery, the wormwood and the gall" (Lam. iii. 19). In Psalm lxix. 21, it is said, "They gave me also gall for my meat." In Deuteronomy xxxii. 32, allusion is made to "grapes of gall," and bitter clusters, as if the fruit of the plant were succulent like grapes, and grew in clusters. On this account some have supposed that it was a species of *solanum,* such as *S. nigrum.* The word *rosh* seems to refer to a poisonous plant, and some have supposed it to be a species of poppy. One species, *Papaver arenarium,* is very common in the fields of Palestine. Celsius thinks that by the hemlock plant is meant the *Conium maculatum* of botanists, the *cicuta* of the Romans. This plant belongs to the natural order Umbelliferæ. It has marked poisonous qualities. This opinion of Celsius seems to be founded on very slender data. This plant seems rather to have been marked for its bitterness than for its poisonous properties. The word *rosh* is in the New Testament rendered by the Greek word *chole,* meaning gall or bile: "They gave him vinegar to drink, mingled with gall" (Matt. xxvii. 34). Again, in Mark xv. 23, in place of gall the word myrrh is used, as indicating bitterness: "And they gave him to drink wine mingled with myrrh." Hengstenberg says in regard to these two passages: "Matthew, in his usual way, refers to theological views in his narrative of the drink. Always keeping his eye on the prophecies of the Old Testament, he speaks of vinegar and gall for the purpose of rendering the fulfilment of the passage in the Psalms (Ps. lxix. 21) more manifest. Mark again, according to his usual way, looks rather at the outward quality of the drink. It was, according to him, wine mingled with myrrh, the usual drink of malefactors." There appears, then, to be no data by which we can determine the exact meaning of the term *rosh.*

# WORMWOOD.

### (PROBABLY A SPECIES OF ARTEMISIA.)

EVERAL species of artemisia grow in Palestine. The common wormwood is *Artemisia Absinthium*. The name always implies wormwood or something more bitter than gall. In Deuteronomy xxix. 18, it is said, "Lest there should be among you a root that beareth gall and wormwood;" and its unpleasant character is alluded to in many parts of the Bible (Prov. v. 4; Amos v. 7; Jer. ix. 15, and xxiii. 15; Lam. iii. 15, 19). The genus artemisia belongs to the natural order Compositæ.

WORMWOOD.—(*Artemisia Absinthium.*)

# BITTER HERBS.

"With bitter herbs they shall eat it."—Ex. xii. 8;
Num. ix. 11.

THE Hebrew word *merorim* has been translated "bitter herbs." It occurs in the two passages referred to above, in which directions are given to the Israelites as to the mode of eating the Passover. They were directed to eat the paschal lamb with unleavened bread and bitter herbs. In Lamentations iii. 15 the same word is used, and is translated "bitterness." In this passage it is associated with wormwood.

There were a considerable number of bitter herbs used along with food, chiefly belonging to the cruciferous and composite orders of plants. Among these are included cresses, lettuce, endive, chicory, elecampane. "Dr. Geddes thinks that the wild endive, or succory, is intended, the leaves of which may be eaten as a spring salad, and the roots dried and made into bread." (*The Scripture Garden Walk*, p. 41.)

The Greek word for "bitter herbs" is *picsides*, and the name has been applied by some to *Picris echioides*, also a composite plant.

## CONCLUSION.

E have thus endeavoured to give a condensed account of the trees, shrubs, and herbs mentioned in Scripture, so far as they can be determined by reference to the best scientific authorities. Although much has been done of late in removing doubts, there are still many difficulties which can only be solved by careful botanical and philological inquiries in Eastern countries.

What an interesting field does the Holy Land present to the Christian man of science, and how valuable might his researches be in throwing light on our version of the Bible. True it is that in regard to the grand truths of salvation he that runs may read, and that the unlearned, under the guidance of God's Spirit, will find the inspired Word profitable for doctrine, for reproof, for correction, and for instruction in righteousness. But there are hidden treasures, the beauty of which is fully displayed only to the enlightened student who applies all the resources of science to their elucidation. We cannot too deeply investigate the words of the Bible, written as they were by holy men who were moved by the Holy

Ghost. Unless we have verbal inspiration we have nothing. The Book not merely contains a revelation from God, but it is, in its words and minutest details, even to the very plants, given by Him. The highest scientific talent may well be consecrated to the noble task of illustrating the Natural History of the Bible.

The unsettled state of Palestine renders the examination of its flora a matter of some risk, and hence there is not an opportunity for leisurely and carefully noting the plants which grow on spots hallowed by associations of the deepest interest. We may look forward to a time when that country, now so desolate, shall be again inhabited, and when its fertile soil will bring forth a luxuriant vegetation; when "instead of the thorn shall come up the fir tree, and instead of the brier shall come up the myrtle tree: and it shall be to the Lord for a name, for an everlasting sign that shall not be cut off" (Isa. lv. 13).